Veloce *Classic Reprint* Series

Brian Moylan's tales of the BMC/BL Works Rally Department from 1955 to 1979

Works Rally Mechanic

Brian Moylan

VELOCE PUBLISHING
THE PUBLISHER OF FINE AUTOMOTIVE BOOKS

Introduction

When I asked Bill Lane, my foreman in the Service Department of the MG Car Company, for a more interesting, possibly more lucrative, job, I had no idea that I was to enter a world of foreign travel, working twenty-four hours a day, following rally cars around the most inaccessible parts of Europe, Asia, Africa and South America.

It was a world that included breaks in some of the most luxurious hotels in the sought-after locations of those continents. And so us raw Berkshire lads had to learn not only to turn our skills to building cars that were to compete with the best in the world, but also the ways and tastes of a society outside our working-class upbringing.

My story is divided into chapters which tell of events that occurred during the eras of the seven managers under whom I worked during my twenty-four years in the department:

Marcus Chambers, who did most to educate us in the ways and delights of gourmet restaurants.

Stuart Turner, who did more than any other man to turn the image of rallying as a drive down to Monte for the rich blades seeking the sun in cold January into one of professional competition between the major motor manufacturers of the world.

Peter Browning, with a hard act to follow, displayed a genius for organisation with his network of support on trans-continental events.

Basil Wales, who picked up the pieces, carrying the burden of representing British Leyland's involvement in motorsport, with a budget largely generated by his Special Tuning Department after 'Competitions' had closed.

Richard Seth-Smith, brought in to boost British Leyland's public image.

Bill Price, who served as assistant to three Competitions Managers, left the company when the department was closed, and returned when Leyland ST rejoined the world of international rallying.

John Davenport, employed as the BL Motorsport Supremo, who presided over the final days of a dynasty which had lasted for over a quarter of a century.

Brian Moylan

Veloce *Classic Reprint* Series

**Brian Moylan's tales of the BMC/BL Works Rally
Department from 1955 to 1979**

Works Rally Mechanic

Also from Veloce Publishing:

www.veloce.co.uk

First published under ISBN 1-874105-97-0 in 1998, and reprinted in 2004 under ISBN 1-904788-18-1 by Veloce Publishing Limited, Veloce House, Parkway Farm Business Park, Middle Farm Way, Poundbury, Dorchester DT1 3AR, England. Tel +44 (0)1305 260068 / Fax 01305 250479 / e-mail info@veloce.co.uk / web www.veloce.co.uk or www.velocebooks.com.
Reprinted April 2018. ISBN: 978-1-787113-30-5 UPC: 6-36847-01330-1.

Contents

1

Marcus (Chubb) Chambers

Looking back at the foundation of the British Motor Corporation Department in 1955, one must not forget that the MG Competitions Department had been revived, after a long sleep, in the autumn of 1954. It was a time when I found myself with much to do and few facilities with which to get started.

However, my greatest asset turned out to be a nucleus of Berkshire men who were both skilled and loyal to the marque. John Thornley, my boss, started me off with some fine material, and they knew where to find more of the best men in the factory, so by early 1955 we were well on the way to forming a team which was to become a legend in motor sport.

One should remember that nothing can be achieved without the skill and enthusiasm of the mechanics who build the competition cars. Those mechanics who attend the events have been motivated by their belief that the cars and their drivers are capable of winning. Support in the field means working long hours in far from ideal conditions; often they have to cover nearly as many miles as the competitors in order to reach the service point on time.

I know that my drivers had regard for the Abingdon boys and were always helped on by the knowledge that their cars would be checked over as well as possible in the time available - and they were sent on their way with a smile.

Marcus Chambers

Marcus Chambers

The Racing Shop

'Brian, we want somebody to go down to Nice to fettle up the Monte recce cars ready for them to start recceing again after Christmas. You'll go tomorrow and come back on Christmas Eve. You OK to do that?'

This was typical of the sort of request or instruction that came your way if you were a mechanic in the BMC Competition Department, as I had been since 1955. That was the year the British Motor Corporation, with a good deal of prodding from MG Managing Director, John Thornley, re-formed the old MG racing shop.

A nucleus of men with pre-war experience of racing at Brooklands was available, still using their expertise on the vari-

The Racing Shop, closed by Lord Nuffield in 1935. This early racer is one of the C Types that took the first five places in the 1931 Double Twelve Hour race at Brooklands. (Courtesy Triple M Register MG Car Club)

ous 'undercover' racing activities in the Development and Experimental Department. Alec Houns-low had once ridden as riding mechanic to the great 1930s Grand Prix driver Tazio Nuvolari, and was now foreman in the Development Shop. His chargehand was Henry Stone, another of the 'old school'. Added to these were the men experienced in building the record-breaking cars: Harold Wiggins, Jimmy Cox, Cliff Bray, Brian Hillier and Gerald Wiffen (who had started in the department as tea-boy but had proved himself such a useful team member that he was kept on as a fitter when he 'came of age').

These, then, were the men already in position, but development work had to continue and the staff needed to be augmented by skilled men recruited from other parts of the factory. Doug Watts came from the Rectification Department and Tommy Wellman from Service Repair.

At this time I was working in the Rectification Department, on loan from Service, where I had been employed as a skilled fitter some five years earlier. Nobby Clark - another Service fitter - had been drafted into the Racing team, but the long hours necessary in the job were not to his liking, and I took his place on my return from Rectification, while he went back to Service.

Marcus Chambers, fresh from the ill-fated government-inspired African Groundnuts Scheme, was brought in as Competition Manager. Marcus had

Pre-war racing mechanics Henry Stone, Bob Scott, Reg Jackson and Alec Hounslow. Henry and Alec were the nucleus of the team when the Racing Shop re-opened in 1955. (Courtesy Don Stone)

Henry Stone, racing mechanic. (Courtesy Don Stone)

much racing and rallying experience and was a well-travelled man, especially on the continent of Europe: he knew every worthwhile restaurant from Calais to Monte Carlo. Evidence of that was his substantial figure (which gave rise to his nickname of Chubb). Dick Green from Aston Martin came with him to be the Racing Shop foreman.

The new department's first venture into racing was the 1955 Le Mans 24-hour race, with three prototype MGAs. Alec Hounslow and Dick Green were in joint command of the other mechanics - Doug Watts, Jimmy Cox, Cliff Bray and Gerald Wiffen - travelling to the race. Besides the six drivers and their wives, the rest of the party comprised various timekeepers, technicians, journalists and advisers, numbering fifteen in all.

A car transporter had been bought and kitted out for its role as mobile workshop, sleeping accommodation and catering unit (when not actually transporting a car). The transporter was driven by Dickie Green, with Gerald Wiffen and Harold Wiggins as passengers. The four cars were driven by Jimmy Cox, Doug Watts, Alec Hounslow and Cliff Bray. Jimmy had a fright on the way to Dover when his car spun out of control and shot up a bank - fortunately resulting in no more damage than a bent numberplate.

The Team's headquarters were in the *Château Chene de Coeur*, courtesy of Captain George Eyston's friendship with the Lady of the Château, Comtesse De Vautibault. Here, the stabling provided an excellent workspace, and the forecourt was adapted for use as the training area for the drivers to perfect their Le Mans start technique of running across the track, leaping into the car, starting up

Right, top: Alec Hounslow and Jimmy Cox standing to the left of Phil Hill, who drove the record-breaking EX 181 to 254.91mph in 1959. (Courtesy Jimmy Cox)

Right, below: Harold Wiggins, Dickie Green, Doug Watts, Alec Hounslow, Syd Enever, Jimmy Cox, Cliff Bray and Gerald Wiffen with the 1955 Le Mans MGAs and Transporter, in the factory yard. (Courtesy Marcus Chambers)

and getting away in good position.

The mechanics also came in for their share of training: bonnet unstrapped for an under-bonnet inspection and oil top-up, wheel change, lights cleaned, refuel and go!

During the first night's practice, the car driven by Johnny Lockett threw a fan belt and badly overheated, but appeared to have suffered no further damage. Nevertheless, the mechanics lifted the cylinder head and gave the engine a thorough inspection. The car of Teddy Lund/Hans Waeffler, MG no. 64, was a reserve for the race but gained its place when the Arnott spun off and was wrecked. This, though unfortunate for the Arnott team, gave the MG team an advantage: they were able to spread into the adjacent Arnott pit, now vacant, relieving the overcrowding they had been suffering in the double pit space the three MGs had been sharing with the Macklin Austin-Healey.

The MGs all got away smoothly from the race start and their progress continued to the schedule they had set themselves. At approximately 5.30pm the first driver change was due. Jimmy Cox was anxiously looking out for the overdue Dick Jacobs from his vantage point at the pit counter - from where he witnessed the subsequent appalling accident and the sequence of events leading up to it. The Macklin Austin-Healey was approaching, full of fuel, having

already 'pitted'. Mike Hawthorn, in the D-Type Jaguar, overtook and dived in front of the Healey, having left until late his entry to the pits. Macklin was forced to take evasive action and moved over into the path of Levegh's Mercedes - which rode up over the Healey's rear wing, taking off into the bank. Here, it disintegrated, throwing lethal debris into the crowded grandstand and killing 87 people.

Amid the ensuing turmoil the whereabouts of Dick Jacobs was still unknown. It was not until nearly 8.00pm that he was reported to be 'slightly' hurt and in hospital, after he had crashed into the embankment at Whitehouse Corner and been thrown out of the car. It was soon obvious that his injuries were more than 'slight', and after two days in hospital he contracted pneumonia. BMC were appealed to for help, and very soon a medical team arrived from England and Jacobs was transferred to the Churchill Hospital in Oxford. This was the end of Dick Jacobs' career as a driver, but he continued his association with the sport and MGs by preparing and managing teams of MGs for the next nine years.

The remaining two Le Mans cars finished the race, averaging 86.17 and 81.97mph for the twelve hours, and were placed fifth and sixth in class.

In 1995 I was one of a group of MG Car Club members who drove our cars to Le Mans to see the race. Whilst there we visited

Dirk Jacobs, car number 42, after his accident at Le Mans.
(Courtesy Marcus Chambers)

the *Château Chene de Coeur*, where we met the Comtesse de Vautibault, by then a frail 92 year old. She was very moved by the memories that our visit evoked and the photographs of the oc-

casion that I had with me.

I join the Competition Department

By the time that the cars had returned from Le Mans I was a member of the Racing Department and was put to work preparing one of the racers for the Dundrod TT in Northern Ireland. With the car completely overhauled, arrangements were put in hand for testing at the Motor Industry Research Association (MIRA) with the drivers and mechanics. On the Saturday before this event my wife was coming to the end of her term of pregnancy, but she insisted that nothing was going to happen that day and that I should go into work as usual. I lived just ten minutes away from the factory and always came home for lunch.

When I arrived home that day I found my wife in labour and the midwife in attendance. Needless

Works Rally Mechanic

Marcus Chambers talking to Nancy Mitchell before the start of the 1956 Monte Carlo Rally. (Courtesy Marcus Chambers)

to say, I did not return to work that afternoon. But my wife was being well looked after, so I went to work on the Sunday. The baby's arrival was not accepted by Alec as sufficient reason to absent myself the previous day, and my commitment to the cause was under grave suspicion all the time that I worked under him.

Two of the cars had with twin-cam engines, one a Morris and the other an Austin that Tommy Wellman had fitted. Tom was in bed asleep at one o'clock the morning before he was due to travel with the cars to Dundrod, when a knock came on his door. It was Alec. 'Come on', he said,

'you and Henry have to take that Austin twin-cam engine out of the race car and put in a push-rod engine.' No explanation was ever given and the engine was never seen again. Tom says that, in testing, it was far superior to the Morris twin-cam that was used in production.

The only MG finisher in the race was the push-rod-engined car driven by Jack Fairman/Peter Wilson. The twin-cam retired after several pit stops to change plugs, and the experimental aluminium petrol tank split, putting out the Lund/Stoop car.

But the vital part of the Dundrod story was that a pile-up caused the death of two drivers early in the race, followed later by a further accident that claimed

the life of another driver. This tragic occurrence, so soon after the Le Mans disaster, caused the BMC Competition Committee to have second thoughts about sportscar racing; henceforth, the accent should be on international rallying. That being the case, the short-lived Racing Department was renamed the BMC Competition Department.

It was not until three years later that the twin-cam was announced. In July 1958, four of the cars were taken down to the FVRE at Chobham, where they were to be test driven round the two-mile test track by the gentlemen of the motoring press. It was a full scale presentation with lavish hospitality, a military brass band and other cars for the press to play with whilst waiting a turn at the wheel of a twin-cam.

I was one of the mechanics sent to look after the cars. It was a long day and, to break the monotony, we took to driving the spare cars round the track. I drove 'Granny', Pat Moss' favourite motor car, the Morris Minor that she had cut her teeth on in her early days with BMC. After a couple of laps I grew more adventurous ... until I put a wheel over the edge of the track on a corner, into the gravel, which pulled the car into the concrete posts marking the track edge. I got out and sat on the bank dismally contemplating my return to the factory the next day, when I was sure I would be dismissed.

However, back at the control centre I found that my escapade was very small beer. Our foreman, Doug Hamlin, had taken one of the bandsmen for a circuit in a twin-cam and written it off, putting the bandsman in hospital. We were very fortunate at having to suffer no worse than a very torrid time in the Works Manager's office back in Abingdon.

International rallying

My first trip abroad was in 1956 to the Tulip Rally. As an exercise in Rally Support this must truly be regarded as an utter farce. Tommy Wellman, Gerald Wiffen and myself, complete with the transporter, went across to Holland with Marcus Chambers.

Scrutineering saw the exclusion of the lightweight MG Magnette with its aluminium body panels that had been built for Nancy Mitchell to drive in her quest for Ladies Championship points. (Understandably, Nancy was not happy at being given a car that did not comply with the regulations.)

The evening before the rally was not spent poring over maps and marking service points, because there weren't going to be any! Instead, we spent the evening having dinner in the luxurious Huis Ter Duin hotel, where Rally Headquarters were established, before moving on much later to our own hotel, the Verloop, just down the road. Here we found a convivial party in

progress with *Autosport* editor Gregor Grant in the chair doing justice to a bottle of Dutch gin, while regaling us with his wartime tales. 'Were you a Company Commander then?' I foolishly asked. Gregor drew himself up, 'I, sir,' he thundered, 'was Colonel-in-chief of the regiment!' *Whoops!* We left the party later on and retired to bed, only to find when we awoke next morning that Gregor - who was competing in the rally - hadn't bothered with bed. When the rally started a couple of hours later, there he was - immaculately turned out and looking as fresh as a daisy behind the wheel of the Lagonda that was his rally mount. Once

clear of
the crowds, however,
Gregor handed over the wheel to

his co-driver. His rally was short-lived, nonetheless, as they soon crashed into some roadworks.

After the rally started it quickly cross the border, but not so us. The transporter was not the vehicle to go charging around the French Alps in pursuit of rally cars. So we were scheduled to stay in Holland until the cars returned three days later. Meanwhile, we visited the beautiful tulip centre, the Kuekenhof, and the model dockyard, the Madurodam. It was during this time that I was introduced by Marcus to the art of fine living (an excellent tutor!)

The rally cars returned to Holland, the final checkpoint being in the Phillips factory in Eindhoven. It was announced that the winner was

Raymond Brookes, BMC's first rally winner, seen here with John Gott before the start of the Liege-Rome-Liege Rally, 1957, in which they came 9th. (Courtesy Marcus Chambers)

Works Rally Mechanic

Pat Moss and Ann Wisdom with their beloved Morris Minor. (Courtesy Den Green)

Den Green fixing the rally plate to the Gott/Brookes MGA Twin Cam. (Courtesy Den Green)

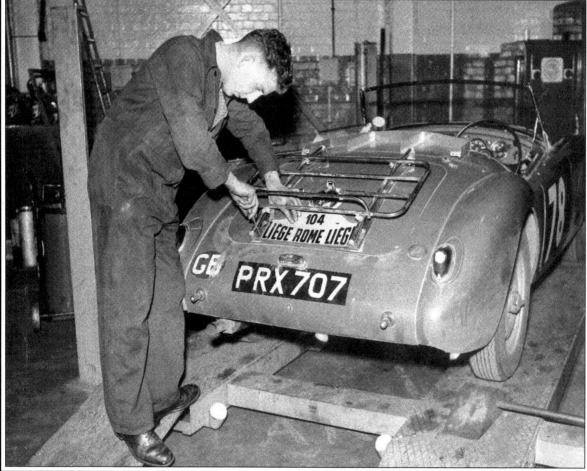

Raymond Brookes, with his father as navigator, in their Austin A30. A BMC winner! Tom was dispatched by Marcus to offer his services, to be told by Brookes senior that 'they had completed the rally without any assistance from BMC and they didn't want any now, so bugger off!'

However, over dinner - and with the aid of a few bottles of the best wine in the cellar - Marcus Chambers smoothed out the ruffled feathers and offered them a works drive. The Brookes team was to compete in four rallies in works cars, and Raymond Brookes went on to become an accomplished co-driver, sharing an MGA with John Gott on several events, the most notable being a ninth place in the Liege-Rome-Liege in a twin-cam.

Parting of the ways

By 1957 it had become obvious that Marcus and Alec were not happy under the same roof. It was at this time that the Service Department was reduced and moved to the main factory; half of the staff were deployed to other parts of the factory and the remainder moved out of the large workshop to a part of the main factory, where they later became the newly-formed Special Tuning Department. Marcus and his team of mechanics took over the vacant workshop in B1. Doug Watts was made foreman, with Tommy Wellman his deputy.

Sestriere 1957

Later that year I went with Marcus to the Sestriere Rally in Italy, where the team comprised Nancy Mitchell with Anne Hall in the MG Magnette, Jack Sears and Ken Best in an Austin A105 and the Brookes duo in a Morris Minor. We saw the cars off from the start and, with the rally underway, set off to our first service point in a Morris Oxford Traveller, accompanied by the motoring journalist Joe Lowrey. Our route took us to within striking distance of San Marino, a principality, whose main source of income was from the sale of postage stamps. The design of these changed regularly, making a visit mandatory for an avid collector like Marcus. We duly called in to this tiny city on the top of a mountain and Marcus made some interesting purchases. Time, then, to catch up with the rally.

With Marcus' foot flat on the floor, the Oxford built up a good head of steam and we were hurtling down a long straight road that disappeared into the distance. Ahead of us were two cars and we were gaining on them rapidly. Just when we were committed to overtake, the rear car moved out to overtake the one in front, completely blocking the road. Marcus braked hard and escaped up the inside where, alas, there was a tree which leapt out and smashed our right-hand wing. Luckily, most of our speed had been taken off, thanks to Marcus' quick reactions, so, although the car was undrivable, the only injury was to Joe Lowery who hit his nose on the wind-screen, while I was taking cover behind the seats.

So, after attending two international rallies as a back-up mechanic, I had yet to wield a spanner in anger.

Accident on the Oxford by-pass

In 1957 two MG racing drivers, Robin Carnegie and Dick Fitzwilliam, bought four MGAs and formed a racing team. The cars were being prepared in the factory for a race at Nurburgring. I prepared one car with Doug Hamblin. As always seemed to be the case, the cars were only just made ready on the day before they were due to be shipped out. That day happened to be a Sunday; that didn't signify much, Alec Hounslow didn't believe in days off and Sunday was just another day.

Amazing though it seems now, the final high-speed test was made on a public road. The straightest and longest stretch in the vicinity of Abingdon was the Oxford Southern by-pass, which, in those days, was a three-lane road. We would drive to the crest of a slight rise, from where we had a good view of the whole stretch of the road, then, when it was clear, we would start the run. With the diff. ratio that was fitted we should reach 6800rpm before the road junction at the far end.

The only other vehicle in sight was an Austin A 70 parked near the far end of the run on our side of the road. Doug was driving and we were edging up to the maximum revs that the car should be capable of - his foot was flat on the floor. The A 70 driver indicated his intention to pull out. We were still far enough away for me to press the central horn button to warn him of our approach; we needed just that extra length of run to avoid having to start all over again. But the A 70 driver not only pulled out, he started to make a U-turn, blocking the road. Doug braked and pulled to the nearside. The MGA went broadside on, into the back offside corner of the Austin - and I went out over the windscreen. I must have rolled when I hit the path some ten yards up the road, for I was able to jump straight to my feet. Doug was sitting quite still, staring straight ahead. I feared the worst, but when I reached him he shook himself and said that only his wrist was hurt (it later proved to be broken). The driver and passengers of the A70 were unscathed - but were a very funny colour.

The MGA chassis was bent but the engine and running gear were undamaged. Somehow, the car was got back to Abingdon, where Alec said 'We've still got to get four cars ready for tomorrow morning, so you'd better start getting the engine and the rest of it into the ex-Le Mans car.'

The next day, to play safe, I reported to the surgery and explained about the accident and being thrown out of the car. The works doctor, Doc White, asked me if I had broken anything, No I hadn't, Did I have any pains? No

again. 'What are you doing down here then?' He said, 'You'd better get back to work!'

Racing with a Riley 1. 5

Les Leston was a Grand Prix driver who was spoken of in the same context as the likes of Stirling Moss, Graham Hill, Roy Salvadori and the rest of the front-line drivers of the 1950s.

After winning the 1954 500cc Championship in a Cooper, Les went on to drive for the Connaught team and later for Aston Martin, for whom he drove at Le Mans in partnership with Roy Salvadori. He was an early driver for BRM, helping to iron out the bugs that beset the car in its early days. He was also the second driver for the Austin-Healey involved in the tragic Le Mans accident of 1955.

During his racing career Les had a set of overalls made which he had fireproofed, and these laid the foundation for the Nomex brand of racing overalls widely used by competition drivers from that time on. There proved to be a ready market for the overalls and other items of racing regalia, which Les sold from a shop that he opened in High Holborn.

At the end of 1957 Les decided to retire from big-time racing and concentrate on running his business, but he still wanted to compete at weekends, so he bought a Riley 1.5 for racing and rallying. Les did a deal with Marcus Chambers whereby the car was prepared at Abingdon and a mechanic was provided to give

support at race meetings. I was that mechanic.

We gave the engine the full treatment: everything lightened and balanced, special crank, racing cam, raised compression ratio, polished and gas-flowed head, and MG exhaust manifold. The body was stripped of all unnecessary trim, and I cut one-and-a-half-inch holes in all the interior body panels. Les would come down sometimes during the week and cut a few holes for himself where he thought I had left too much metal.

Our first outing with the Riley was to Brands Hatch, Les' 'home' track where he had a permanent stall in the paddock to sell his go-faster bits. Here he was the crowd's favourite son and he delighted them by breaking the class lap record and bringing his Riley 1.5 home in third place behind Tommy Sopwith driving a Jaguar and Jack Sears in his Austin A110.

Incidentally, Sears and Sopwith tied for the Saloon Car Championship that year and the decider was a race at Brands Hatch with both of them driving identical Riley 1.5s for five laps, then changing cars for another five laps. When it came to change-over time, Jack Sears said to me in a stage whisper: 'Where is that secret switch you promised me?' Tommy looked nonplussed for a while, then it dawned on him, 'Of course, you're a BMC works driver, aren't you Jack?' He said. I don't think he was ever sure if it was a joke or not - especially as Jack Sears ran out the winner of the contest.

Later on we went to Aintree, where the scrutineers took exception to the Riley's MG exhaust manifold. Luckily, Peter Reece, who had had some BMC works drives, was made aware of the problem and he took us to his BMC garage in Liverpool, where I took the manifold off a Riley in the showroom window and borrowed it for the race. But we were beaten on this occasion by Alan Foster in his modified MG Magnette, prepared by Dick Jacobs.

At the 1958 BRDC *Daily Express* Trophy meeting at Silverstone, the Saloon Car race was dominated by the exciting duel for the lead between the three Jaguar drivers - Mike Hawthorn, Tommy Sopwith and Ron Flockhart. But our interest was in the renewal of the contest with Alan Foster, who was lying sixth, one place ahead of Les (who had another MG Magnette chasing him). This all changed when Alan Foster lost a wheel at Becketts and the other Magnette spun off, leaving Les to win the class - and, with the Rileys of Harold Grace and Peter Taylor, we also took the team prize.

In the same year Les entered the Monte Carlo Rally and took with him Paddy Hopkirk. This was Paddy's first major rally, and for him it was a harsh introduction. They were to compete in the race-trim Riley, and Les - with his obsession with weight-saving - refused to have even the

The author with Les Leston.
(Author photo)

heater back in the car. Instead, he bought a little paraffin heater that he hung off the dashboard. Little did he know that this was destined to be one of the harshest Montes on record: Les - along with the eleven-strong BMC official works team - failed to reach Monte Carlo.

Monte Carlo 1959
My earliest recollection of the

Monte Carlo Rally is the 1959 event. In 1957 the rally had been cancelled due to the Suez crisis, and in 1958 the weather was so atrocious that no BMC works cars finished. John Gott's car for that event was a Wolseley 1500, and I had been involved in some continental testing with the car.

The trip was made memorable for me by a meal in the Hotel Renaissance in Saulieu. The culinary delights of oyster and snail were, I fear, a little wasted on my

Above: John Gott, Doug Watts and the author with the Wollesley 1500, which was being tested prior to the 1958 Monte Carlo Rally.

Left: RAC Rally 1963 - before studded tyres. Doug Watts fits chains for the Morely brothers.

Left: Alan Foster, on the left, and Syd Evener and son, Roger, with a Dick Jacobs MG 1100. (Courtesy Den Green)

untutored palate but the rest of the meal - which culminated in a tour of the hotel's extensive wine cellar - was a milestone along my road to appreciation of how the other half lived.

Preparation of cars for the Monte, in those days, was concerned more with reliability and comfort than outright speed. Sheer speed was only possible when the advent of spiked snow tyres virtually allowed the drivers to ignore the ice and snow, secure in the knowledge that their tyres would grip and take the car where they pointed it. For this particular event the tyres used were Dunlop Weathermasters, with strap-on chains for when the going got really tough.

Crew comfort was important because it could get very chilly a couple of thousand feet up an Alp in January. So heaters were modified with extra-efficient matrices fitted in them. Keeping the windscreen clear of ice was a problem which we attempted to solve by fixing a perspex strip four inches wide across the top of the dashboard, keeping the air from the demisters against the screen; electric bar windscreen defrosters were also fitted.

Reclining seats were a luxury item not found in BMC cars of that period, and we were instructed to 'make the passenger's seat fold back so that the crew can rest when not driving.'

We were not given any more specification than that and were left to do the job in the best way we could. Ron Bettles, one of the fitters who joined the department when Competitions broke away from Development, came up with a device which incorporated the use of the long screw part of a wind-up jack. The seat-frame was cut at the join of the back to the seat and bolted together to form a hinge, a bar was welded across the bottom of the seat-back with the nut from the jack in the middle of it. The screw with its

One of the first Works Minis in the 1959 RAC Rally, driven by 'Tish Ozanne'.

handle was fixed to the front of the seat for the occupant to wind the back up or down.

Halda Speed Pilots were fitted: these clever little instruments told the navigator whether he was running faster or slower than his intended average speed. The Halda had two dials - one was a clock, the other an average speed indicator. If the intention was to travel at an average of eighty

kilometres an hour, for example, then the hand on the indicator dial would be turned to eighty. The clock face had an additional hand behind the minute hand. If the car was achieving a higher average than indicated, the additional hand would creep in front of the minute hand and vice versa. This device was run off a cable from a T-piece on the back of the speedo and had to be calibrated to a particular car. As well as the dials, there was also an odometer, which was adjusted by means of a screw.

We had a 9.6 kilometre test circuit that we would drive round to achieve accurate calibration. The 9.6 had to click over exactly when we regained the start point; if it didn't, a turn on the screw would adjust it in the desired direction, and off we'd go, round again. This was a painstaking and time-consuming operation, made more tedious by the steady speed at which we had to drive, sticking as close as possible to the same line each time. Keeping to the average speed could be a deciding factor in the Monte -with secret checks set up along the road to monitor competitors' averages.

The rally had many different starting points. Paris was a popular one, as there were no further ferries or borders to cross (and, of course, there were the attractions of the place itself). Our cars were kept in the garage of the Paris importer, De Fries, where we worked on the final preparation. The draughty workshop was heated by a stove on which a kettle full of red wine (which was constantly replenished) was kept hot all day. With work over for the day, we were taken by the foreman and mechanics of the garage to their favourite local bar. A lively cou-

a nightclub and went on to Fred Paine's bar in the Rue Pigalle. Fred was an expatriate Englishman who - I have subsequently discovered - served

Bob Whittington, Gerald Wiffen, the author, Tommy Wellman, Ron Bettles and Nobby Hall in the Transporter. (Author photo)

stayed to see all the cars safely away. Ernie Giles and myself were to make up a service crew with Marcus. After the cars had all left we retired to bed for a couple of hours, as we were not going to catch up with them again for some time; other crews had been dispatched earlier and would rendezvous with our team members before we saw them again.

About nine o'clock the following morning we were on our way down the N7, complete with a picnic

Den Green with his while hair patch. A legacy from two experiences as a rally navigator.

p l e   of hours followed, during which the postman arrived and was encouraged to take a drink with us ... and another ... and another ... until, finally, he threw his bag of mail over the bar, said 'Poof! - I'll deliver that tomorrow', and promptly fell asleep. That night we visited

with the French underground during the war.

The rally start was the following evening and went on well into the night, our last car not getting away until the early hours of the morning. Marcus hamper per which had been provided by the De Vreis garage proprietor (this included, incidentally, a fruit knife, which I still have thirty-eight years later).

After a night spent in a Chubby Chambers Recommended hostelry, we eventually caught up with the rally at Bourgoin on the Chambery road from Lyon. No emergencies awaited us. Pat Moss had suffered a broken inlet manifold on the A40 but this had been efficiently repaired in a local garage *en route*, so, after carrying out routine servicing on all of our entries, we made our way back to the main drag down to Monte.

We arrived in Monte in good time to meet the cars and give them all a thorough checkover to ensure they were fit to tackle the 450 kilometre mountain circuit that night. Our working headquarters in Monte Carlo were in the Garage Sporting, proprietor M. Jacquin. After working we would be invited into the flat above the garage, where Madame Jacquin would feed us on onion soup. I have never tasted onion soup anywhere that could compare with Mme Jacquin's. It was here that Marcus announced he intended to be out on the mountain circuit that night and asked for a volunteer to go with him. Nobody seemed very keen, but I thought it would be a great experience and so put my hand up. It was an exciting night, spent dashing round the mountain roads behind Monaco, occasionally meeting up with the cars,

whose drivers seldom had time for us to do anything except clean the windscreen and headlights.

It was a successful rally by our standards at that time. Pat Moss won the Ladies' prize and was tenth overall. Out of the six entries, John Gott (twin-cam) and Raymond Baxter (Austin-Healey Sprite), were the only non-finishers.

Den Green - navigator in Portugal

Servicing was not allowed on the Portuguese Rally of 1959. Marcus had entered four cars in the event: an Austin-Healey for Pat Moss, two Minis to be driven by Peter Riley and Nancy Mitchell, and a Wolseley 6/99 which he was to drive himself, with Den Green, to provide back-up to the team.

The journey through France was punctuated by stops at various hostelries. Pat Moss and Ann Wisdom left their passports in the last of these before crossing into Spain. They drove the Austin-Healey back to collect them and were hurrying to catch up again when they were in collision with a taxi. Marcus and the rest of the team were having a meal when they had the phone call from Pat telling them of the accident. He and Den took off back up the road immediately to see what had to be done.

The front end of the Healey was pretty extensively damaged: the front wings, headlights and the radiator were smashed. Den worked all that night to repair the damage and the next day they

drove to the start of the rally in Portugal.

Eric Carlsson, who was driving a Saab with John Sprinzel, had injured his back in an earlier rally, and Den was asked to bind him in tape from his armpits to his waist to ease the pain.

The route that the rally took was frequently crossed by railway lines. At one of these, Carlsson tried to beat the barrier as it came down, but it struck the car which rolled on to its roof, where it lay beside the track with the crew inside it while the train went roaring past. Despite the added pain he was in, Eric righted the car, which was still drivable apart from the stove-in roof. The roof was beaten out and an emergency screen fitted at the next stop. They went on to win the event!

Marcus and Dennis started the rally without having had any sleep since before they repaired the Austin-Healey. They were stopped at one of the level crossings in the early hours of the morning, where they both fell fast asleep, waking in broad daylight with not another rally car in sight. A look at the road book told them where they could expect the rally to be. With Den reading the map, they took some short cuts and, in time, found themselves travelling behind Pat Moss. Den was following the route on his map and directed Marcus to turn right, but the Pat Moss Healey had gone straight on and Marcus, putting his faith in the navigational skills of Ann

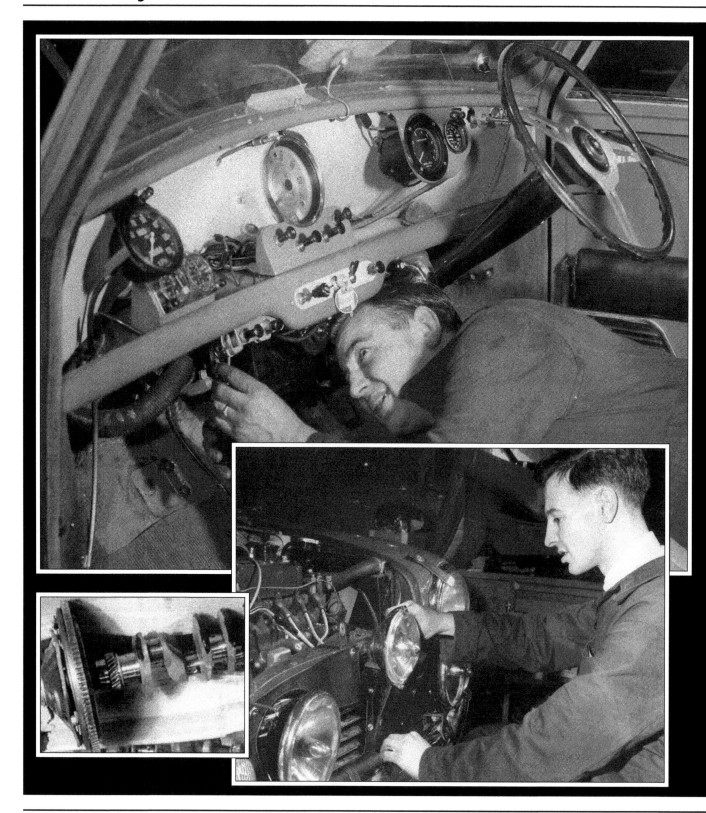

Wisdom, followed it. Soon the big rally car had left them behind and disappeared from sight, only to come screaming back as the occupants realised they had taken the wrong turning ...

Pat and Ann won the Ladies' award and, as a token of their appreciation for his hard work, they gave Den the trophies they had won, which - along with the Finishers Plaque awarded to him and Marcus - still has pride of place among his souvenirs.

Den Green - navigator in Norway

The 1960 Viking Rally in Norway was another event that banned organised service, and Marcus used the same ploy as he had in Portugal: entering the rally to provide back-up. Den was again his navigator.

The route wound its way through the mountainous region of Norway and Marcus struggled to keep up in the heavily-laden Wolseley. The hairpin bends followed quickly on each other and, finally, Marcus lost control of the car, which headed toward the edge of the road. Den was sure they were going to go over, but they hit a large rock which threw

Main pic, left: Ex-prize fighter, Johnny Organ, finds working under an 850 Mini dash very cramped. (Courtesy Den Green)

Far left: The primary gear situated behind the flywheel allowed oil on to the clutch.

Left: Roy Brown fitting the Rad Grille Muff, first used on the 1960 Monte Carlo Rally. (Courtesy Den Green)

them back against the mountain wall. Still continuing on its way down the mountain, the car scraped along the wall, with sparks flying and the smell of paint burning off the bodywork. Eventually, they were able to stop, and Marcus - congratulating himself on having got away with it - prepared to drive on, but Den impressed on him the need to assess the damage before tackling the rest of the mountain road. They were amazed to find that the outer skin of the passenger's door had been completely ripped off, revealing the door catch and window-winding mechanism. The front wheel was damaged and, where they had scraped along the wall, the wheel studs were ground-off level with the nuts.

However, they continued servicing the Pat Moss A40 and again she won the Ladies' prize. Den's memento from this experience was a patch of hair that turned white.

The Stockholm start 1960

Another early Monte was the 1960 event. This rally was significant in being the Mini's first Monte Carlo Rally. The Mini had made its rally debut on the Mini Miglia National Rally late in 1959, with Pat Moss and Stuart Turner taking the car to a win on its first outing.

The second appearance was a three-car team on the RAC Rally, but the car was showing some serious teething troubles which resulted in all three failing

to finish. One of these troubles was a slipping clutch, caused by oil leaking from the oil seal in the primary gear. This gear slides over the end of the crankshaft and is externally splined to be driven by the clutch plate. When the clutch is disengaged, it revolves independently on the crankshaft and therefore needs lubricating. To this end, the crankshaft had an oilway drilling with an oil seal in the gear to prevent the oil from oozing out of it on to the clutch plate, causing the clutch to slip. It didn't work. Despite modifications to the oil seal, oil still leaked through. The answer was to design a bush in the gear that did not require lubrication. Gears fitted with these bushes were made available only after three of the Minis had left Abingdon for their Monte Carlo Rally start in Stockholm.

I was given the task of going to Stockholm, fitting the new gears and blocking off the oilway. The work was to be carried out in the workshop of the Swedish BMC Importer. I was allocated a local fitter to assist me and, by four o'clock that afternoon, we had taken out two of the Mini engines, changed the gears and put the engines back in again. The workshop closed at four-thirty and, to my astonishment, all work ceased at four; the men went off and washed their hands and then sat on their benches waiting to knock-off. And this was at a time when the British motor industry workers were receiving a bad press back home!

Nevertheless, two of the cars were done. I didn't know or care very much whose cars they were, but the drivers paying a visit to check on progress found that the one remaining to be done was Nancy Mitchell's car. For this I was roundly castigated by Nancy who would not be convinced that I hadn't left her car 'till last because 'I suppose you don't think a woman's car is as important as the men's.' In fact, Nancy's was the only car of the three that didn't finish - but none of them suffered with clutch-slip.

Another spin-off from my trip to Stockholm was the introduction of the rad-grille muff, which was made and marketed in Sweden, as a necessary extra. Our method of trying to raise the water temperature had been to use a radiator blind, which was fitted between the radiator and the outlet vent under the nearside front wing. This was a spring-loaded roller blind, cable-operated by the driver. It was not at all successful because the blast of air coming through the front passed over the engine on its way to the radiator, so the heat wasn't being generated in the first place. I took some of these muffs down to Paris where I was to join !he rest of the team and they were fitted to the Paris start Minis, greatly adding to the drivers' comfort.

Bill Price

In October 1960, Bill Price joined the department, fresh from National Service after an apprenticeship in the Nuffield Organisation. Bill was to be an assistant to the Manager. Over the years the job grew in importance, and when Marcus moved on Bill provided continuity as Stuart Turner's right-hand man, a role he filled again as Stuart was succeeded by Peter Browning.

One of Bill's major responsibilities was to register with the FIA every detail of the BMC cars, including parts produced in sufficient numbers to be permitted for use in modifications to the cars. This information was the car's 'homologation', and at rally scrutineering the car would be checked against its homologation to satisfy the scrutineer that no illegal parts had been fitted.

One of Bill's unofficial jobs was to help with organisation of the annual parties, a task which Bill threw himself into with great enthusiasm.

Acropolis 1961: cocktails in the Astir Beach Hotel.
(Courtesy Paul Easter)

The majestic Corinth Canal.
(Courtesy Paul Easter)

The Acropolis Rally 1961

My first visit to Greece. Our party for the journey included Donald Morley, Ann Wisdom, David Seigle-Morris, Vic Elford, David

Hiam, Mike Hughes, and Tony Ambrose's wife Barbara, besides Marcus and myself. The rally cars with other crews were going by a different route.

Just getting there was an experience, stopping *en route* for two nights in northern Italy before reaching Venice, where we spent a morning sightseeing, riding in a gondola and visiting a glass works. Later, we were to embark on the boat that was to take us down the Adriatic to our destination.

We approached the Greek port of Pireus through the Corinth Canal in the early morning. I was sharing a cabin with Donald Morley and we both went on deck to take in the majesty of the canal, whose sides towered above us. The journey was cautiously made with a tug towing the boat and very little space on either side.

We disembarked later that morning and, after completing customs formalities with the cars, made our way to the Astir Beach hotel on the outskirts of Athens in Glyffada. Here, we were accommodated in luxury chalets. Early in the evening a waiter would come round to take orders for dinner. T awards the hour when the meal would be delivered, a maid appeared to set the table on the veranda. The meal arrived by tricycle - equipped with a hot-box - and with sufficient wine to last the night out. Usually the mechanics would all eat at the same chalet, visiting a different one each night.

But there was work to be done also. As always at the start of a rally, the crews would test their cars. This invariably resulted in changes to specifications - harder or softer shock absorbers, seat positions altered, lights refocused. This particular operation was carried out at night with a mechanic accompanying the driver over a typical part of the rally course and setting the lights precisely how the individual driver wanted them. They concentrated especially on the long-range driving lights which would be set inclined towards each other in order to give light round an approaching bend at the earliest possible time; there is nothing worse than a beam of light stretching straight out over an abyss and no road in sight.

The cars were driven quite a large part of the way to a rally start and faults could be found en route. They always had a complete pre-rally oil change, and the hyped-up crews' nervous foibles had to be catered for. Finally, a wash and polish to present the team cars in pristine condition at scrutineering. So some full days had to be spent in the local BMC workshop.

In Athens this was the garage of the Doucas Brothers, where we were shown the remarkable coach construction operation. Chassis, engines and drivetrains were imported from Great Britain; the bodywork was hand-produced in the Doucas workshop. The top corners of the coach body were beaten into shape on

a huge hollowed-out tree stump. All the bright-work was made and plated. In one corner a heap of seat frames had been made and were awaiting the upholsterer. The finished product was a luxury long-distance coach.

Among the skilled craftsmen in the garage was a machinist and John Sprinzel needed his services at the end of the rally after he clouted a rock and smashed a brake disc. Doucas already had ingots, roughly cast in the required shape, and in a day a serviceable disc was made ready to be fitted.

The rally brought plenty of work. Our entry consisted of an Austin-Healey for Peter Riley and Tony Ambrose and three Minis. We also had to look out for the other two BMC cars that had been entered privately. One of these was an Austin-Healey driven by John Sprinzel, the other was another Mini driven by ex-Dunlop Competition Manager, David Hiam, partnered by Mike Hughes.

The Greek roads were covered with a film of marble dust and pebbles which made driving a hazardous affair at rally speed -and, sure enough, the accidents started to happen. Vic Elford, driving the Mini that he was sharing with David Seigle-Morris, came up against an immovable rock which tore the tie-bar bracket from the front of the subframe. We were not carrying welding equipment and the repair was painstakingly carried out by a local blacksmith - alas,

Works Rally Mechanic

Above: BMC mechanic's visit to Fort Dunlop in 1960. Top, l-r: two members of Dunlop management, Ernie Giles, Den Green, Nobby Hall, Dunlop tyre fitter. Bottom, l-r: Bob Whittington, Neville Challis, David Hiam (Dunlop) Tommy Wellman and Roy Brown. (Courtesy Mick Legg)
Right, l-r: Dunlop tyre fitter, Tommy Eales, Johnny Lay, Cliff Humphries, Ron Edwards, the author. Bottom, l-r: Gerald Wiffen, Doug Hamblin, David Hiam, Doug Watts, Dunlop rep.

not quickly enough to allow the car to reach the next control before the latest booking-in time, so it was duly disqualified.

There was some chafing at the length of time the job had taken, but on a later Tour de France the same thing happened and our ace welder, Nobby Hall, was unable to effect a permanent repair in the time available. The bracket came adrift twice more before the car retired. The ultimate cure, for subsequent events, was to weld additional webs to the bracket where it joined the sub-frame. These webs were shaped so that they also provided a skid for additional protection.

The lack of welding equipment was proving something of an embarrassment. It was needed on other occasions during the rally, notably on Johnny Sprinzel's broken accelerator pedal which resulted in him having to rig up a hand-operated accelerator cable until he could get a repair in a local garage.

The Mini's weak points were still making themselves known. The front shock absorber brackets were breaking. The shock absorber fixing eye was located by sliding it over a pin standing out from the side of the bracket. This meant that the pin was only fixed to the bracket at one end, and under severe conditions it snapped off. Before the Mini's next event this fault was eradicated by extending the bracket so that there was support at the outer end of the existing pin. This made it impossible to slide the shock absorber eye over the pin, and so the pin was removed and replaced by a bolt and nut.

However, a broken shock absorber bracket was responsible for putting the second works Mini, of Mike Sutcliff and Derick Astle, out of the rally. But Donald Morley/Anne Wisdom's exit was much more spectacular. They shot off the edge of a cliff, tumbling end over end before finally coming to rest over 100 feet down. Miraculously, both were unhurt.

Following the rally through Greece took us to some wild countryside, well off the tourist routes, which was an aspect of the job that I particularly enjoyed. Finding a meal could bring about some uproarious incidents. A 'cafe' in a remote village had nothing on offer that tickled our taste buds, but we got through to the proprietor that an omelette

Vic Elford, rally navigator turned Grand Prix driver.
(Courtesy Den Green)

quired a taste for when accompanied by a glass of Metaxa, the Greek brandy, but on its own it's not at all thirst-quenching. 'Nescafe' is the universal substitute for ground coffee - but here in the mountains of Greece they had never heard of it. By now the villagers had crowded into the cafe to gawp at us strange beings. Our request for Nescafe was passed from one to the other with much nodding and shaking of heads, none of which produced the desired result ... so we had a glass of beer.

Later on we witnessed a cafe proprietor approach a woman in a bus queue. She had a kid goat in a basket and was apparently on her way to market. The cafe man bought the goat, hung it on a tree outside his cafe, cut its throat and skinned it. Then he got a hosepipe and sprayed the road to lay the dust and keep the evening meal clean for his patrons.

The rally came to an end with Peter Riley/Tony Ambrose winning the GT category and coming third overall. David Hiam and Mike Hughes were second in their class behind Eric Carlsson, who was outright rally winner.

Vic Elford, driving the rally Mini, missed the boat home. Fortunately, the boat had one more port of call on the Greek mainland and he was able to catch up, in spite of no money and having to barter for petrol with his spare cans of Castrol oil.

This was the last outing for Vic in a BMC car, which was a

would do us very nicely. Six of us sat down to eat. Myself and another mechanic were in one support car, Marcus and Doug Hamblin in another, and we were joined by David Seigle-Morris and Vic Elford - who were providing additional back-up after their retirement from the rally. The omelette arrived. I say omelette (in the singular) for that is what it was: one six-egg omelette on one plate around which were arranged six forks. It was every man for himself. Coffee, it was decided, would help it down well. The Greek or Turkish version was all that was available, in minuscule cups half filled with sludge. This is a beverage that I have ac-

Austin-Healeys in production in the MG factory in Abingdon. (Author photo)

shame for he went on to become an accomplished driver, winning the Monte Carlo Rally in a Porsche. Later he turned to Grand Prix racing, competing at Nurburgring and at Brands Hatch in the 1968 British Grand Prix at the wheel of a Cooper Formula One car.

Austin-Healeys

Undoubtedly, the rally car of this era was the Austin-Healey. A car with no finesse, just a big engine, an unbreakable frame and suspension, although it didn't start out like that.

The car had been in production in the Austin works at Longbridge, Birmingham since 1952. It had been jealously guarded by Austin (now BMC) boss Leonard Lord and had been the reason why MG was not allowed to produce a modern sportscar - until in 1955 Lord gave way and permitted production of the MGA. In 1957 production of the Austin-Healey was moved to Abingdon, where the

car was quickly taken up by the Competition Department.

The engine used by Austin was a well-tried 2.6 litre unit. Five Austin-Healeys were entered for the Alpine Rally of 1958 and final placings of the four finishers were seventh, tenth, eleventh and twelfth. Jack Sears, who was to top the team's placings, made a good start on the speed test at the Monza Autodrome, putting up second fastest time to a 300SL Mercedes. He went on to clock up fastest time of the day on two other circuits. Pat Moss' car was almost put out of the rally by a well-meaning journalist who, seeing the rocker cover breather hanging loose after the gauze air filters had been removed, tidily tied it in a knot. The effect of this was to allow a build-up of crankcase pressure which forced oil through the rear main oil seal and on to the clutch, causing so much clutch slip that the car could hardly make it up hill. Johnny Organ was the one to spot this error and, with the crankcase able to breathe again, the car was soon back up to speed - allowing Pat to finish tenth and win the Ladies' prize.

A car that didn't finish was John Gott's. One of the splined flanges that bolted on to the rear hubs to carry the knock-on wheels sheared, allowing the wheel to go rolling off over the edge of the mountain and the Healey almost followed it! The piece of flange still on the hub was inspected by Terry Mitchell, and on his recommendation the flange was modified for subsequent use on all the production cars.

Terry from the design office was a great character. He had built himself two specials and a miniature railway that ran round his garden. At one point on the rally we were servicing opposite a team from a continental manufacturer. This team was apparently highly organised, with four vans and a car carrying the *chef d'equipe*. When a car arrived this *chef d'equipe* would go to the car,

Timo Mackinen tackling a flooded road on the 1966 Scottish Rally. (Courtesy Den Green)

ask the driver what problems he had and then relay the information to the mechanics with a loud hailer, whereupon the appropriate action would be taken.

Unfortunately, the group had no welding equipment. A car pulled in with a broken dynamo bracket and, in great Gallic agitation, the *chef d'equipe* came to ask if he could use our equipment to weld it. At first we said that he would have to wait until all of our cars had gone through, but we relented and one of the mechanics started the repair. It was a heavy bracket and the mechanic wasn't getting the job hot enough; consequently, he was having great difficulty and making a very ugly job of it. This so upset perfectionist Terry that he took the torch off the mechanic and did the job for him.

Jumping and boiling

1959 saw the introduction of the Austin-Healey 3000 with its three two inch, twin-choke SU carburettors. Later on in the quest for more power, these were replaced by twin-choke Webers which showed a significant improvement. Doug Hamblin, the deputy foreman (who had been elevated to that position when Doug Watts moved up to supervisor and Tom to foreman), was sent to the Weber factory in Bologna, with Gerald Wiffen and a Weber-equipped Healey. The Italian experts soon decided on the optimum jet settings and other refinements. Doug and Gerald came back with a useful knowledge of tuning the carburettors - which none of us had had dealings with before.

More power meant more

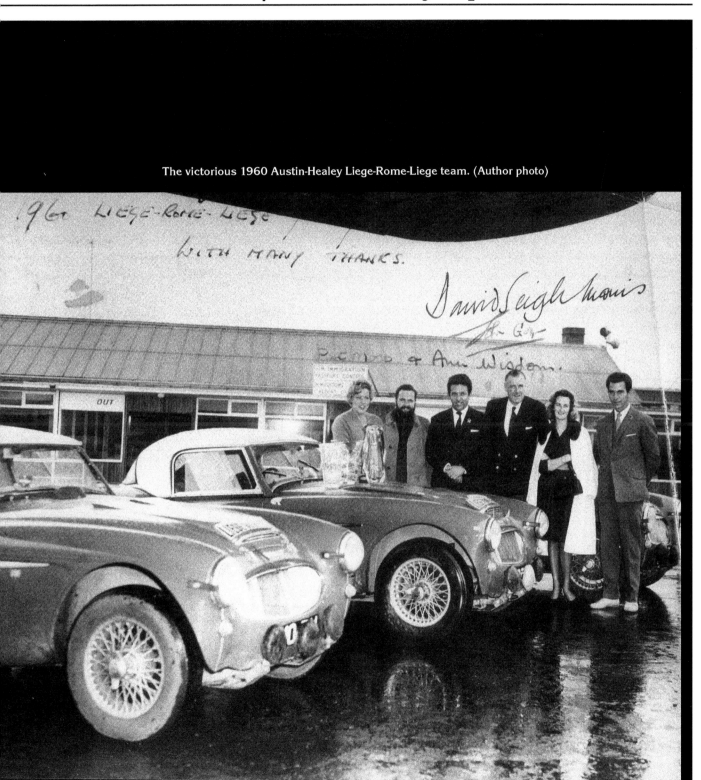

The victorious 1960 Austin-Healey Liege-Rome-Liege team. (Author photo)

strain on the transmission and gearbox failures were reduced by fitting straight-cut gears.

Further back the semi-elliptic 'cart springs' developed a tendency to tear loose from the front mounting and appear under the seat. A test run without the rear parcel shelf fitted allowed us to see the dreadful distortion of the springs when power was applied. The torque reaction of the axle caused the nose of the diff to twist up, making the springs the shape of the letter 'S'.

Bars mounted in brackets welded to the top of the axle, going forward to the frame parallel to the spring, cured this but the chassis was failing where the axle ran across it. The axle was fitted over the chassis frame, limiting the amount of movement of the axle and causing the axle to crash down on to the chassis member on full rebound. Another inch of movement was gained by cutting this amount from the top of the chassis, replacing the lost strength with a couple of lengths of angle iron, and using stronger springs.

The strength of the springs was tested on a particularly sharp, humped-back bridge on a back road. I remember Doug Hamblin driving the car while I hung over the back of the seat observing the performance as we took off from the top of the bridge and landed some distance along the road. The bridge was over the river Windrush by a riverside pub whose patrons would line the road exhorting us to further

efforts to see how far we could make the Healey jump!

Up at the front end the lever-arm shock absorbers had to be prevented from tearing free from their platform by welding additional plates to the mountings. These 'shockers' were losing efficiency when worked hard as the fluid in them boiled. We overcame this by mounting a canister above them, filled with shock absorber fluid, which was connected to the top of the 'shocker' by a tube. The extra capacity of fluid reduced the problem, and topping up this canister was one of the routine jobs at a service point.

Marcus' swan-song

If the Austin-Healey was the rally car of the era, then the event for which it was bred was the Liege-Rome-Liege Rally. Known also as the Marathon de la Route, this rally evolved into the most fearsome event of its kind when tourist road conditions in Italy forced it off the Italian roads in 1956 and down to Zagreb. Later, the rally travelled the whole length of Yugoslavia into Bulgaria and back again to Liege. As well as the almost non-existent roads in Yugoslavia there were the fearsome passes of the French and Italian Alps, raced over on the way down and on the way back!

But in 1960 the Healey was in its earlier stages of development. Nevertheless,, that year's Liege gave Marcus the biggest win of his career as BMC Competition Manager. Pat Moss won the event

outright, despite having a gearbox change during the rally. The cars driven by David Seigle-Morris and John Gott were fifth and tenth respectively, to secure the Manufacturer's Team Prize, and they took the first three places in their class. To add to the Healey domination, John Sprinzel was third overall and first in class in the Sprite.

So Marcus was able to go out at the top; the following year he resigned, handing over the reins at a time when the team was just beginning to sweep all before it. His swan-song was on the 1961 Liege-Rome-Liege, and his successor, Stuart Turner, went along to get acquainted with his new team, under combat conditions.

That things were going to change was made plain when Stuart rode in a service car with Doug Hamblin on Marcus' last event. Their first rendezvous with the cars was down on the Yugoslavian Autoput, after the rally had already undergone some gruelling mountain tests. I was in another service car with Tommy Wellman at the same control and remember Stuart chafing at the bit and vowing never to be so out of touch with the rally again.

After our stop, Tom and I were scheduled to be at the final service point on the Italian side of a border crossing, in order to deal with any problems that the harsh Yugoslavian roads had caused the cars, and to prepare them for the final part of the rally that would take them back to Liege. The Healeys of Don Grimshaw and John Gott had both been put out of the rally; there remained the cars of David Seigle-Morris and Pat Moss. Seigle-Morris came through and continued to the end of the rally to take first place in the class, but after a long wait it was obvious that Pat was not going to arrive and we eventually got confirmation that the suspension - that later we would extensively modify - had broken and the car was undrivable.

I went to the border crossing post with Tom and we asked to be allowed to recross the border to retrieve the car and, more importantly, the crew. Our request was turned down. In order to enter Yugoslavia it was necessary to have a visa which could only be used once and we had used ours. The border official told us that to re-enter Yugoslavia we would have to go to the embassy in Trieste and apply for another visa. Trieste was some distance away, and we decided to return to Liege secure in the knowledge that the team instructions had covered such an eventuality: Marcus and a mechanic were sweeping up the route and would surely bring them home. Alas, it was not to be. Marcus, finding the going too tough for his service car, had left the route and opted to take the better main road and get back to Liege as soon as possible. At a meeting to discuss the situation that evening it was decided that John Gott should go to the rescue because he still had a re-entry visa. A collection of Yugoslavian dinars was made and John - accompanied by Eric Carlsson, the Swedish rally ace who became Pat's husband - left that night to find the girls, who had been stranded for four days in a remote village behind the iron curtain.

An argument broke out at breakfast next morning, as Marcus accused Tom and myself of not making more effort to get across the border. I retaliated by pointing out that we didn't need to because he was sweeping the route. After that no more was said on the subject.

Stuart Turner

At the time that Marcus Chambers announced his intention to resign from his post as BMC Competition Manager, I was a sports editor for Motoring News, and to be suddenly catapulted from that to running the BMC Competitions Department was a dream. The timing couldn't have been better. The Austin-Healey team had just swept the board on the Liege and this was only in the car's early stages of development. The Mini had yet to make its mark, but I was well aware of its potential, having navigated for Pat Moss to win the Mini Miglia Rally in an 850cc version back in 1959.

The RAC Rally of 1962 was my first as a Competition Manager, and I spent it rushing around the route in an Austin Al JO, which was the unlikely but sensible choice of car for carrying all the required paraphernalia for rally servicing. The car was driven by Austin-Healey ace, Peter Riley, and Brian was in the back to deal with emergency repairs. We carried a fluorescent octagon sign and set it up wherever we stopped, giving the impression that BMC service was everywhere.

The workshop at Abingdon was supervised by men of great experience. The mechanics were all skilled fitters who had been creamed off from other departments in the factory, and the drivers were the top of their profession.

The successes that followed

Stuart Turner

were achieved by a combination of the above ingredients; in other words, 'teamwork'. The blanket servicing was hard work for all concerned, but the abiding memory of my period of tenure is just what a marvellous team it was. Great days!

Stuart Turner

Rally servicing Stuart Turner style

An early taste of Stuart's brand of rally servicing was experienced on the 1961 RAC Rally. He followed the whole rally in an Austin Westminster with Peter Riley (driving) and myself to cope with emergency repairs, as we visited places where the other service crews couldn't reach. On one occasion we found Donald Morley with a punctured tyre, running on his spare. We took the punctured tyre, drove to the nearest Dunlop service van to collect a new one and took off to catch Donald again as soon as possible. I was in the back of the Westminster with a pair of

Pete Bartram and the author concentrate on the engine, whilst Gordon Pettinger of Dunlop tightens the wheelnut on Pat Moss' class-winning Austin-Healey in the 1961 RAC Rally. (Courtesy Den Green)

tyre levers, struggling to get the punctured tyre off the wheel. I couldn't refit the tyre in the back seat. We caught up with the rally and, with the rally car following, made for a nearby garage where I was able to refit the tyre and get it blown up to pressure.

Donald had previously damaged his sump, causing considerable oil leakage, so while we had the car in the garage he drove up on to the ramp and I brazed the small split in the sump. I had drained the oil from the sump but, during the brazing operation, there was a tremendous explosion of the gases in the engine that nearly gave me heart failure. A plus, however, was that it blew out some of the dents in the sump and didn't seem to have any detrimental effect! (Donald and Erle Morley eventually retired with a broken rear hub bearing that lost them too much time before they could get it repaired.)

My other abiding memory of the event was travelling along a gated road and leaping out of the car every few minutes, opening and shutting gates.

This was the first outing for the recently MG-badged Midget, and MG achieved second place in the Manufacturers' team award. Derek Astle and Mike Sutcliffe were respectively first and

The author tuning the Tommy Gold MG Midget in preparation for the 1961 RAC Rally. (Author photo)

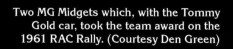

Two MG Midgets which, with the Tommy Gold car, took the team award on the 1961 RAC Rally. (Courtesy Den Green)

A snow covered 'frog-eyed' Austin-Healey Sprite on the 1962 Monte Carlo Rally.

The Reverend Rupert Jones on his way to a third in class on the 1962 Monte Carlo Rally. (Courtesy Den Green)

Donald and Erle Morley's class-winning MGA coupé being prepared by Johnny Organ for the 1962 Monte Carlo Rally. (Courtesy Den Green)

second in class, whilst Tommy Gold, driving his own car, made up the team, finishing well down the list after a lengthy repair to his front suspension following an accident. Pat Moss in a Healey took second place, won her class and the Ladies' award. David Siegle-Morris, also in a Healey, came second in class to Pat and was fifth overall. Altogether a satisfactory first time out for our new leader.

Monte 1962

The Mini was by now beginning to show its potential. John Cooper had been running a team of racing Minis and Morris Engines Development Department in Courthouse Green, Coventry, had modified an engine for use, to great effect, in Cooper's Minis. The practice of using a successful Grand Prix racing team's name on the car gave rise to the Mini Cooper.

This was to be the Mini Cooper's first Monte Carlo Rally. Two of them were entered: one a full factory car with the Pat Moss/Ann Wisdom crew, and the other a car owned by Geoff Mabbs and driven by Rauno Aaltonen, the first Finn in a BMC car.

Rauno was driving on the Mountain Circuit, after the rally had reached Monte Carlo, when he hit an outcrop of rock, dislodging the secondary fuel tank and turning the car end over end; then it caught fire. Rauno was unconscious and Geoff Mabbs, unable to get out of his door which was jammed against

a wall, had to climb over him to reach the driver's door before he could free Rauno from his harness and drag him to safety. The harness had a strap that passed through a buckle with a quick-release clip. The fire had already reached the strap, turning the nylon end into a ball that would not release back through the buckle. Heroically, Geoff tore Rauno from the harness and out of the car. The inverted car burned so fiercely that the carburettors were left as shapeless lumps of aluminium.

Valuable lessons were learned from this near-tragedy and, as a result, this type of safety harness was discontinued. How the secondary fuel tank was secured was also modified. This tank was fitted on the right-hand side of the car and was on the opposite to the standard tank. The tank is secured by means of a strap that runs from top to bottom round the tank, following the line of a ridge. The strap on this secondary tank ran to the outboard side of the ridge. When the back end of the car hit the rock, it was able to push the tank out of its strap on to the battery, causing an immense short circuit and consequent inferno. Positioning of the strap was altered to run round on the inboard side of the ridge on subsequent cars and the battery fitted with an insulating cover.

To Yugoslavia with 'Mombasa Mac'

Stuart was really getting into his

stride when Liege time came round again. The rally now travelled down the length of Yugoslavia to Sofia, the capital of Bulgaria, and the BMC cars had blanket coverage the whole way, with teams of mechanics leapfrogging to keep up with the team. This was always a breakneck affair as, before pushing on to be in position before the first car reached the next service point, the service crew had to wait for the last car in the team to go through.

Petrol stations were few and far between and the fuel dispensed was of a very low octane rating. There was also the problem of queues. The sparsity of filling stations meant that the whole rally filled up at the same time and cars that were way back in the queue would lose valuable time. Added to this was the chance that if you were towards the end of the queue, it was possible that by the time it was your turn there was no fuel left! We carried fuel in containers on the service 'barges', filling up before we left Italy, with the potent-sounding *Supercortemaggiore*, but there was a limit to the amount we could carry. This was to be Bill Price's first rally and he was given the job of driving a 'barge' specially fitted with a Dunlop custom-built rubber 100 gallon petrol tank mounted on the roof. This was filled up in Titograd and eliminated the need for our cars to waste time at filling stations.

I was given a travelling companion from East Africa, 'Mac' Carpenter, a man who had competed in the East African Safari and who regaled us with so many tales from Africa that we dubbed him 'Mombasa Mac'. He took the wheel as we left Abingdon, in order to become accustomed to the way the 'barge' handled, and I soon had my first lesson in how an experienced driver coped with rough roads in Kenya. On approaching one of those high ridges in the road - left after road workers had dug a trench across the carriageway - Mac twitched the wheel to the left, catching me unawares. This move, Mac assured me, was imperative to ensure that the car didn't crash straight over a ridge but, instead, took it at an angle, allowing one wheel at a time to cross. And so we proceeded, slewing across the road at every obstacle, much to my consternation and that of all the other road users (other road users were not a consideration in Africa - there were not too many of them).

My next lesson was how to take a hump in the road. The theory was that, as the front hit the top of the hump, it was thrown into the air. To counteract this, one should give a sudden dab on the brakes to throw the weight of the car forward; then, as the rear wheels reached the top, a squirt on the accelerator would throw the weight back again. Mac took every opportunity he could to teach me the East African Safari driving techniques. Consequently, I did more than my fair share of the driving ...

We travelled in the company of Bill, with his tank-equipped 'barge', and Johnny Organ. John was a colourful character, an ex-professional boxer who had fought for the Southern Area Championship. He had learned his trade as a motor mechanic at Oxford's City Motors, along with many other skilled fitters who came to MG lured by the higher wages and the chance get into the Competitions Department. (Den Green, who was later to be made foreman of the department, was another City Motors product).

Johnny Organ ran on a fairly short fuse: he had been asked to leave his last job after throwing a starter motor at the foreman. His other claim to fame was that he grew a gourd on his windowsill and wrote a very successful book on the subject of gourd growing. He was driving a barge and towing a trailer loaded with spares - including a vast quantity of spare wheels.

Initially, the route in Yugoslavia was without problems and we spent the first night in one of the tourist hotels that were being built at the start of Marshall Tito's endeavour to capture some western currency from visitors to the superb Adriatic coastal resorts.

That night we had a meal in the hotel restaurant, helped down by the retsina wine of the country (an acquired taste if ever there was one!) and finished off with generous glasses of Slibovitch, the Yugoslavian plumb brandy.

I staggered off to bed ... but for Mac the evening was just beginning and, before retiring, he got rid of another bottle of Slibovitch with precious little help from anyone else. I was amazed the next morning to see him up early and going for a shower, dressed in his night attire - a colourful sarong tied round his waist, which he assured me was what all red-blooded males wore in Kenya.

We set off for Titograd. The coast road was the only one marked in red on the map, but it was nonetheless difficult to follow. It was constructed of rocks a little larger than cricket balls and generally had a little less vegetation growing on it than had the road verges. This was not always an infallible guide, however, and often it was a case of choosing the less evil-looking stretch of terrain so long as it headed in our general direction.

Little wonder, then, that our heavily-loaded barges were soon showing signs of distress. Johnny Organ, with his trailer load, was not coping at all well and, at one point, took the desperate measure of jettisoning over a sea wall into the Adriatic some of what he considered to be the least important items. This included some of the spare wheels, and he was quite disconcerted to see them float away in line astern down the coast. Even with the lightened load the barge was finding it heavy-going, and eventually the tow hook broke. Bill Price hitched the trailer on to his car, where it started to shake the front suspension to bits. However, with half a dozen mechanics present we managed to get the cars to Dubrovnik safely, and there we parted company with the Johnny Organ barge, leaving him to wait for the cars to arrive a couple of days later.

Meanwhile, I accompanied Bill to Titograd, where he was to fill up his rubber tank with 100 gallons of Yugoslavian super grade. Titograd was a stark menacing army garrison town which Mac and I were happy to leave, but an indelible memory of the place is of the women with their dark hollow eyes, mutely holding out a hand for whatever scraps we could spare. Their children clamoured round, some of them barely three years old, begging for cigarettes -which we assumed would be used as a valuable form of currency. In this we were mistaken, for no sooner had we handed some out than they were finding matches and smoking them!

The charcoal burners

Our way now led inland towards the mountains of Montenegro. We left the red road that we had been following when we were sixty miles from Titograd. By now night was falling and, appalling though the road was, there was no mistaking where it was, for to one side was the mountainside and to the other a sheer drop. We picked our way gingerly along this goat track. Close to its highest point, some 1900 metres above sea level, we be-

came aware that there was a fire glowing in a clearing ahead of us. When we reached the clearing we saw that it was a band of charcoal-burners, huddled round their smouldering charcoal mound, an eerie sight in the dead of night when we imagined ourselves to be the only human beings to have ventured that way since time began.

We reached Pee, our destination, in the early hours of the morning, completely exhausted. We gratefully parked and soon fell asleep, waking next morning to the sounds of the village stirring. The notes that had been prepared by the team navigators on their pre-rally reconnaissance, and given to us along with an official rally roadbook, allowed us to pinpoint where the control should be set up. Accordingly, we prepared our service point in a convenient clear area the appropriate distance before the control. When the Rally Controllers arrived to set up their control table they were not as sure as we had been of the correct site, and it was a matter of negotiation to persuade them that our service area was outside the limits of the control. This was always an important factor because it was against rally regulations to service within the control area, but we needed to be as close to the control as possible so that any servicing could continue until quite near to the crew's booking-in time.

Here, we were not intended to be a major service point but

an emergency stop. Even so, the barge had to be unloaded to the extent of having items to hand that we were pretty sure we would need -like brake pads and shoes, oil and water, jack and tools - and to unload the roof rack of spare wheels. We would also connect the welding gear to save valuable seconds if we suddenly needed to use it.

The BMC entry was made up of four Austin-Healey 3000s, two Morris 1100s and an MGA Mk2 coupé. Only two Healeys survived long enough to reach us, along with Pat Moss' Morris 1100 and John Gott in the MGA. They had been racing over the type of roads that we had treated with so much respect, and consequently the cars were shaking themselves to pieces. So it was a matter of jacking them up and going over every nut and bolt with a spanner, checking brakes, topping up oil and water and cleaning off the layer of dust that had settled on the windscreens. The time allowed for the rally was almost impossible to achieve and, consequently, the navigators would be standing over us - stopwatch in hand - eager to clock in.

After seeing our last car depart we cleared up and set off, back along mountain roads we had traversed the previous night. Now it was daylight, and until we regained the main road we were on the rally route with rally cars catching us up and overtaking without hesitation, expecting us somehow to find room to get out of their way.

Mechanics Den Green, Doug Hamblin, Gerald Wiffen, Doug Watts and the author, all decently covered on the French Riviera, with the 1962 Alpine Rally Austin-Healeys. (Author photo)

The going on the main road was not much better than it had been on the mountain road and we were both extremely tired, having snatched only a couple of uncomfortable hours' sleep in the barge the previous night. Mac, who had been telling me at the start of our journey that road conditions in Yugoslavia could not be worse than those in Kenya, was by now revising his opinion and said that although the roads could be just as bad in Africa they were not as bad for as long. In fact, it was 400 kilometres back to Dubrovnik, but sometimes dropping to ten or even five miles an hour to pick our way between potholes (or through them when there was no alternative) made the time taken seem like an eternity.

We made the rendezvous with the other two barges at Dubrovnik. From here, the rally route was pretty much a straight line with no chance of either of our service crews catching up for further servicing. So we followed the road home back through Italy, collecting sundry other cars that had retired from the rally on the way, until we had accumulated quite a sizeable convoy.

The two Healeys driven by Logan Morrison and David Siegle-Morris were the only survivors of the BMC team, finishing second and third in class. The MGA retired with a leaking fuel tank from what was to be its last major rally.

Mechanics error

The Alpine rally, which took place in the Alps of southern France and Italy in the middle of summer, was always a great favourite with us mechanics. The start was in Marseilles, and the

Works Rally Mechanic

BMC team was based, before the start, in the delightful little resort of La Ciota, along the Mediterranean coast. We were able to carry out all our pre-rally work in the pleasant surroundings of the grounds of the *Rose The* and enjoy a cooling dip in the sea when the going got too hot.

But the rally conditions were very severe. The dry roads - even on the high passes where the surf ace was broken up - allowed the cars to be driven at racing speeds, which greatly stressed every component, and brake pads were consumed at an alarming rate. Donald and Erle Morley's Healey retired with a broken experimental limited slip differential. Paddy Hopkirk slid his Cooper 'S' off the road, and the Mini Coopers were falling victim to the front driveshaft rubber couplings which had been the subject of an enormous amount of research by Moultons, the manufacturer. The Mini Cooper would get through three or four sets of these in a rally and, although the experience gained in changing them meant that we could do the job extremely quickly, their failure caused the retirement of at least one car, that of Denise McCluggage. Eventually, the problem was solved by the adaptation of solid Hardy Spicer universal joints.

But it was not a mechanical failure that put out the Mini Cooper driven by Johnny Sprinzel. He was flat out on a downhill stretch when his steering column became detached from the rack. Terry Mitchell, from the MG design office, was in the service car that arrived at the scene first and he immediately noticed that the clampbolt was still in position in the steering column. The splined spigot of the steering rack pinion, to which the column is attached, has groove cut across it. The column has a clamp that the clamp bolt passes through and the column must be pushed down onto the spigot far enough for the clamp bolt to pass through the groove. It is possible for the bolt to go through the clamp before the column is fully home on the spigot and when it is tightened it will feel secure, but a good jerk upwards will disconnect the column from the steering rack. That is what had happened in this case. The consequent loss of steering caused the car to crash into the mountainside. But it would have been much more serious had it gone the other way - where there was nothing to stop the car from disappearing over the edge. Although the car was a write-off, the crew, Johnny Sprinzel and Willy Cave, escaped unscathed.

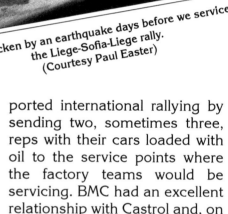

Skopje, stricken by an earthquake days before we serviced the Liege-Sofia-Liege rally.
(Courtesy Paul Easter)

Damaging a Castrol company car

The Castrol Oil Company supported international rallying by sending two, sometimes three, reps with their cars loaded with oil to the service points where the factory teams would be servicing. BMC had an excellent relationship with Castrol and, on occasion, a mechanic would accompany the Castrol rep to form an additional service crew.

And so it was that I joined Ray Simpson in his Ford Zephyr 6 company car for the 1963 Liege-Sofia-Liege Rally. We left the start in Belgium and by lunchtime were in Germany looking for a suitable restaurant in which to eat. We had pulled up on the road outside an attractive-looking Gasthoff where a local, having enjoyed a liquid lunch, was directing his companion's car out of the car park. To our amazement he didn't stop waving his companion back even though it was obvious that he would collide with us. The consequent coming together left a nasty dent in the Zephyr's door, whereupon the merry Germans drove away leaving us dazed and Ray fuming.

Our ultimate destination was Skopje, well down in Yugoslavia and not far from its border with Greece. After a night stop in Belgrade we arrived at our destination, which only a week before had suffered a serious earthquake. The service point was on the edge of the devastated town and the people were living in makeshift accommodation of tents or corrugated iron. The heat was oppressive and after struggling with the clutch slave cylinder under a Healey I almost collapsed. I recovered after dowsing my head under a cold tap, but Ray was having trouble keeping his eyes open, so I took the wheel and Ray dozed beside me. It was a service point we were happy to leave behind.

It was a good road back to Belgrade and we were making good time when there was a crash as the Zephyr bonnet flew open, spreading itself against the windscreen. I was able to stop at the road's edge. The distorted bonnet would not lock down, so I made a hole in it with a screwdriver and held it down with locking wire. From Belgrade we were due at the other end of the Italian Autostrada at the foot of the Vivianne Pass. The police manning the Autostrada start in Trieste were not too happy about the se-

curity of the bonnet and I had to bang another hole in it to secure it with more wire to satisfy them.

During our service stop Paddy arrived to tell us that Rauno Aaltonen had crashed his Healey near the top of the pass. We packed up the car and set out to see what we could do. The Vivianne Pass was not much wider than the Zephyr, and in negotiating a tight hairpin bend on the way up I scraped the side of the car against the mountain. We found the Healey balanced over the edge of the road with its front wheels stuck out in space. A breakdown crew had been alerted but on arrival decided the job would require more equipment than it had and would take many hours to complete the rescue.

Meanwhile, Rauno and his navigator, Tony Ambrose, having almost completed the rally (which they were leading at the time of the crash), were now suffering from reaction to the crash and, feeling exhausted, asked us to waste no more time in getting them on their way back to Belgium.

Ray drove his Zephyr, with our two passengers, back down the mountain and scraped the other side of the car on the same corner as I had done on the way up! Altogether, we had damaged nearly every panel on the company car that Ray was going to have to take back to his boss ...

Winning the Monte

Rallying was latched on to as a

The Col de Vivionne, not wide enough for a Ford Zephyr. (Author photo)

prime vehicle for the publicity boys, with the BMC Mini making big headlines. The Monte was a major sporting event and reports from a BBC team covering it were broadcast every evening. This was also the time when 'Beatlemania' was at its highest and a plot was hatched to make a news story out of the combination of these two popular subjects.

The Beatles were appearing in Paris, from where the Minis were due to start the Monte. Ringo Starr had business in London and was travelling to Paris later than the rest of the group.

The BMC rally team was due to depart from England on a Saturday, but the previous Thursday Stuart Turner came into the workshop and asked me if I was ready to travel that day. A story had been concocted that Ringo had been left behind, and when he arrived in France a rally car would be standing by to speed him into the centre of Paris in time for the concert. I zipped off home - where my efficient and long-suffering wife had all my clothes ready to be packed - and it took little time to put them in the suitcase and away we went. With Stuart driving, we were on our way to Dover.

The car we used was the Mini Cooper S of Pauline Mayman and Val Domleo. Val told me years later that this car was chosen in case we had an accident on the way (the Ladies' prize was deemed the most expendable); perhaps I was chosen to accompany Stuart for the same reason! The road from Abingdon to Henley was shrouded in dense fog. Stuart, being fairly new to the area, said 'You should know this road well - tell me when it's safe to overtake!' Miraculously, we got to Paris in one piece, and next day presented ourselves at Orly airport, where Ringo was photographed getting into the Mini Cooper S to be whisked into

Monte Carlo, 1963: the author during final preparation on the Aaltonen/Ambrose car, number 288, the class winner. Tommy Wellman works on Paddy's car, number 66, the second in class.

Paris. However, in the background of the picture is the Rolls that actually did the job after the photographers had gone!

The publicity value of the Monte at that time was so great that Ford America sent a team of Ford Falcons to compete in the event. Rumour had it that these big cars were experiencing difficulty, while recceing, in getting round the narrow mountain roads at speed, so they were employing a road working gang to widen some of the worst sections! In the event, the Falcons were quicker than the Minis but they couldn't beat their handicap.

It was typical Monte weather; freezing and with snow-covered roads. But this year the Minis were equipped with the rad-grill muffs now being made in the Trim Department in the MG factory. They also had electrically heated windscreens. In fact, 1964 wasn't as cold as the previous year's event, when the ferry was breaking through ice all the way across the Channel!

At the end of the rally we heard the devastating news that Doug Hamblin, the Deputy Foreman, had been killed on his way to Dover to join the event. Road conditions

Suggestion of road widening by the Ford Falcon team on the 1964 Monte Carlo rally. (Courtesy Paul Easter)

Headlamp washers fitted on the 1964 'Monte' cars. (Courtesy Paul Easter)

Beatles drummer Ringo Starr hitching a lift to Paris in 1964.

were such that he and the mechanic travelling with him agreed there was enough packed snow to warrant fitting spiked tyres. The car is unmanageable with spiked tyres on a dry road, however, and before they reached Henley road conditions cleared up. Before they had a chance to change tyres the car slewed across the road into the path of oncoming traffic. In the ensuing collision, Doug died.

This news had a severe

the MGB. Although overshadowed by Paddy's outright win, the drive by Donald Morley was no less of an achievement.

The euphoria generated by these results carried us along and we joined in the celebrations. It was not usual for the mechanics to attend the prize-giving dinner in the *Hotel de Paris*. The speeches were long and pompous and in French (taking up valuable drinking time). Instead, we had our own celebration in

to do in the garage, we would take our mid-morning break in the cafe, which could last until lunchtime. On this occasion our hostess, mindful of the fact that it was a special celebration, laid on a magnificent meal. She was a great joker, and on the cheese board there would always be one that squeaked as you went to cut it, and out of the pickle jar would spring a serpent. On this occasion she was particularly attentive as we sampled the nuts.

Service point on the 1964 Tour de France. (Courtesy Paul Easter)

dampening effect on our celebrations when it was learned that Paddy Hopkirk had won the rally; with other Minis fourth and eighth the T earn prize also came to BMC. This result was added to by the Morley brothers, who won the Grand T curing category in

the cafe across the road from the *Sporting Garage*, where the cars were garaged and we would carry out repairs. The cafe was owned by Madame Pascale, a dear lady who ran the cafe with her daughter (who was engaged to the chef). We were well known by Madame as we used the cafe every time we visited Monaco. Very often, when there was little

The reason for this was revealed when, on opening a walnut, a condom sprang out, at which she ran into the kitchen cackling in delight.

The following evening the official BMC celebratory dinner, attended by no less than Alec Issigonis (whose brainchild was the *cause celebre*), was held in a restaurant called The Pirate.

The author preparing Timo Mackinen's
1963 Monte Carlo Rally Austin-Healey.
(Author's collection)

Another splendid meal and an occasion of much self-congratulation and horseplay. Doug Watts came riding into the restaurant mounted on a donkey ... but it couldn't match our own celebrations of the previous night.

The following day the winning Mini was flown back to England with Paddy and Henry Lid-

don to appear on *Sunday Night at the London Palladium* hosted by Bruce Forsyth. The car was given pride of place at the centre of the revolving stage at the end of the show. Peter Bartram, the mechanic who built the car, travelled with them and was in the London Palladium audience.

This presentation of a rally-winning car can't have been as spectacular as Stuart Turner's own experience, though, when he won the RAC Rally with Eric Carlsson in 1960. Prize-giving was at The Talk Of The Town and the SAAB, complete with Eric and Stuart, popped up through the trapdoor in the middle of the stage.

Tour Auto

The Tour de France Automobile, or - as it was colloquially known in France - the Tour Auto, was an event that went on and on, taking in most of the racetracks and hillclimbs in France, plus the Nurburgring in Germany. And it was to the Nurburgring that Bill Price, Gerald Wiffen and myself headed for our first service point. The works cars were all behaving themselves, but the Mini Cooper, privately entered and driven by Terry Hunter, was burning copious amounts of oil. We removed the cylinder head to see if the problem could be overcome. The fault was caused by too much having been machined off the face of the cylinder head in order to raise the compression ratio, and breaking into the oilway that takes the oil feed up to the rocker gear. This early in the event Terry was loath to retire and begged us to weld up the oilway, which we did, warning him that, without a full oil feed, the rockers wouldn't last long.

Another frailty of the Mini engine showed itself on Timo Mackinen's car when the bottom fan belt pulley came loose and sheered the Woodruff key which located it on the crankshaft. Later investigation showed that the shims behind the bottom timing gear, which were there to line the timing chain up with the camshaft timing gear, were breaking up. The fan belt pulley was tightened against this gear and disintegration of the shims left the pulley loose. In future, the shims would not be used and lining up the two timing gears would be done by machining the boss of the camshaft timing gear.

Another potential cause of the same problem was the lock-washer behind the head of the pulley fixing bolt, which would get hammered flat, again leaving the bolt loose. A special locking washer was made to avoid having anything between the bolt head and the pulley. This was a plate with a double hexagon cut in it to fit, as a ring spanner does, round the bolt head, and it was secured by bolts in the pulley which were in turn wire-locked.

On the track at Rouen the Ford Galaxies, with their big, straight-through exhaust pipes, made the earth shake as they roared through between the grandstand and the pits. But Paddy, in the diminutive Mini Cooper, was not to be shaken off and he passed them on every corner, only to be overtaken again on

Above: Tulip Rally winner 1964: the author servicing the Mini Cooper 'S' of Timo Mackinen on its first ever outing. (Author photo)

Main pic: Pauline Mayman with Val Domleo up above the snow line in June on the Alpine Rally of 1963. (Courtesy Paul Easter)

Inset: The author contemplates the rear axle as the Morley twins look devastated at missing out on a Coupe d'Or. Alpine Rally in 1963. (Author photo)

the straights. At the end of the eight days of racing Paddy was third overall.

At Cognac there was a race on a military airfield. No servicing was allowed but we were able to get ourselves on a corner to spectate. Andrew Hedges, driving the MGB, was by far the most spectacular. We were parked next to a team of Italian mechanics from Alf a Romeo who jumped up and down with excitement, cheering Andrew on and urging him to greater efforts every time he appeared, almost sideways,

Timo Mackinen and Mike Wood at the desolate Italian/Yugoslav border crossing on the Liege-Sofia-Liege rally of 1963. (Courtesy Den Green)

with two wheels on the grass.

The Mini that provided much entertainment at Le Mans was the car driven by John Wadsworth. He was losing and burning oil at such a prodigious rate that on every lap he would stop at the pits, where I was posted with a quart can of oil. I didn't have to dip the sump, just up-end the can into the filler, punch a hole

in the bottom with a screwdriver to allow the can to empty, slam the bonnet shut and wait for him to come round again for another lot.

Tommy Eales christened him 'oily' Wadsworth. His rally was soon to end; he just couldn't carry enough oil to get him from one service point to the next.

The event was sponsored

A line-up of the 1964 Alpine Rally cars in the MG factory compound. (Courtesy Den Green)

by Shell, who provided lavish hospitality at every circuit, with gourmet meals and unlimited amounts of wine (which the drivers were unable to appreciate). This was made up for by some of the mechanics, giving rise to some strange pit signals when races were run after lunch. Decorating the tables at these meals were some simple but attractive table napkins. I acquired a set of six, which are still used on special occasions.

The Alpine *Coupe d'Argent*

The objective on the Alpine Rally was to finish the event without losing any time on the roads or timed tests. Any competitor who accomplished this feat was rewarded with a cup - more properly a *Coupe des Alpes*. If a competitor was to win three of these during his career, he was awarded a silver cup, or *Coupe d'Argent*. If those three wins were in consecutive years then it would be a golden *Coupe d'Or*. Gaining a *Coupe d'Or* was the supreme Alpine accolade - Stirling Moss was one of only three drivers to do this. But to win a silver was almost as rare an achievement: only four of these were awarded throughout the event's history. The only other BMC driver to gain the award was Donald Morley in Austin-Healeys. Donald's wins were in 1961, 1962 and 1964. The broken differential in 1963 robbed him of a gold award. Paddy Hopkirk had won a *Coupe des Alpes* in 1956 and 1959 before joining BMC, and

he was confident that this year, 1964, driving the Mini Cooper S, he would win a coveted silver.

The Mini Cooper Ss were all suffering with a bad batch of rear wheelhubs. New ones were hastily summoned from England and were delivered to us at our service point. I was a member of the service crew with Bill Price and Robin Vokins. Two of the Cooper Ss came in together. I changed the rear hubs on Pauline Mayman's car while Robin got to work on Paddy's. There was no time to lose ... we were in sight of the control that they had to reach on time to keep alive their hopes of a cup. Pauline's car went without a hitch and she made her way to the control. Paddy's hubs were changed and one brake drum refitted; the second one jammed on the brake shoes. The shoes had been adjusted right back and were properly centralised; still the drum wouldn't go on. Paddy saw his precious *Coupe d'Argent* disappearing and urged us on 'Come on boys I'm losing my *Coupe*.' In desperation I took a copper cosh to the drum and forced it on. As soon as Robin had put the wheel back in place and started the wheelnuts Paddy roared off to the control. We relaxed, thinking that he had made it, and hoped that he remembered to tighten the wheelnuts before he started the timed test.

Flaming brakes

Towards the end of the first day of the rally we were scheduled to see the cars in to the night halt in

Cannes. We were high in the Alps and set off on our way down to the coast. On rounding the first bend we came across a broken-down Mini that had been entered by a private owner, David Friswell. There was nothing that we could do to repair his blown engine so we agreed to tow him to the nearest town, where he could get further assistance. The Mini crew pleaded exhaustion and I was volunteered to take the wheel of their car as we hitched it on to a rope and made our way down the mountain. The road plunged steeply down from one hairpin bend to the next. At every one Bill, driving the barge, slowed to negotiate the tight bend. I had no option but to brake as I careered toward his rear bumper. The brake pedal went further and further towards the floorboards as the brakes faded, until - mercifully - we reached level ground at the entrance to the town of Grasse. I frantically sounded the horn. Bill stopped, then got out of the barge in a hurry. 'I've got no brakes left!', I shouted. 'I'm not surprised', he replied, 'they're on fire!' Robin leapt out with the CO_2 extinguisher and quickly doused the flames, but before we could congratulate ourselves they burst into life again. The seals in the front brake callipers had disintegrated in the intense heat and brake fluid was catching fire as it leaked onto the hot discs. We jacked up the car in the middle of the wide road and removed the wheels to play the CO_2 on to the discs to cool them;

then we retired to a nearby bar to cool ourselves and plan our next move. When we judged that it was safe to refit the wheels we set off again, slowly towing the Mini along the road until we arrived at a garage, where we left the car and crew to their own devices.

We arrived at last in Cannes where the rally had beaten us to it. By this time it was late in the evening, and we were horrified to learn that one theory for our absence was that we were ashamed to show our faces because our protracted service on Paddy's car had indeed cost him his *Coupe.*

His disappointment was comparatively short-lived, however, for he achieved his ambition in the next year's event.

Slavonic ballads

In September we were once more in Yugoslavia and I was again teamed up with Tommy Wellman and Paul Easter. Our first service point was beyond Split, in a village called Ogulin at a quarter to two in the morning, with enough petrol to put four gallons in the three Austin-Healeys and the Mini. Our entry in the event also included three MGBs, but our instruction made no mention of them re-

Liege-Sofia-Liege 1964: Altonen and Ambrose leaving the start of the rally, which they won. (Courtesy Den Green)

Timo Mackinen retired from the Liege-Sofia-Liege with punctures. (Courtesy Den Green)

quiring petrol from us at Ogulin. Stuart must have foreseen that they wouldn't reach that far, as indeed they didn't. Two of them succumbed to clutch trouble and the third broke a spring shackle. Timo Makinen also failed to put in an appearance with his Healey,

which had suffered so many punctures he had been unable to reach a Dunlop service van before he ran out of spare wheels!

After Ogulin we were to proceed to Obrovac, where Tom was instructed to drop me off at a hotel near to the control. I think Stuart must have written this instruction with his tongue in his cheek for, after making our way once more into the Yugoslavian outback, we found the village had one cafe and no hotel.

- Tales of the BMC/BL Works Rally Department 1955-1979

The cafe owner, however, could speak a little English, and with that and a lot of sign language we learned that a lady in the village let out a room in her house where I could sleep. The lady in question was a charming person of advanced years who had met English people before and she showed me postcards that they still sent her.

The rally was due through the village the next day, and I spent the rest of the day acquainting myself with the geography of the village, establishing my service point by a fountain where the detailed BMC route notes - written after weeks of recceing - told the crews to zero their speedo trips.

I was agreeably surprised to find a man fluent in English. There was a drink bottling factory where this man was the doorman. During our conversation he voiced his opinion that Winston Churchill was the best politician of all time and that our current Prime Minister, Harold Wilson, was no good at all. Hardly politically correct from that side of the iron curtain! He also very prophetically told me that Yugoslavia was a hot-bed of ethnic discontent which was kept under control by Marshall Tito's leadership and that following his death a blood-bath would engulf the country.

I awoke the next morning to find that the village marketplace was outside my window, and the peasant farmers from the area had arrived that day to sell their produce. It was a scene of great bustle and cheerful noise with the braying of donkeys and bleating of goats mingling with the good-natured banter between the people who obviously met their friends only on these rare occasions. The business transactions took up most of the morning. The men took little part in the proceedings and congregated in the nearby bar; by midday they were singing most harmoniously what I can only describe as Slavonic ballads.

The cars were due through in the afternoon and so I reluctantly left the happy scene. Taking my tools and some oil and water that Tom had left me with, I took up my station, which was on the rally route between controls. I was without spares, other than those carried by the competitors, and was there in case of desperate emergencies, of which there were none. But I felt that my presence was justified when, on checking the brakes on one of the Healeys, I found that the pads were worn thin and might not last to the next control. New brake pads were in the car's kit of spares and I was able to change them. After the routine checks our last car left, assuring me that all the BMC cars left in the rally had passed through my service point.

My next job was to telephone Doug Watts and report to him that all of the cars which had passed through his control at Split had reached me safely. But in this I came up against a problem. There was a Post Office with a telephone, but it would take two hours to get space on the only line connecting Obrovac with Split ... and the Post Office closed in one hour!

I returned to spend the rest of the day with the Slavonic choir in the bar, by now in fine voice, while their womenfolk sat under the trees by the side of the river, enjoying a gossip and counting the results of the day's transactions.

I was picked up the next day, but, before I left, the idyllic image was shattered by the arrival of the police, intent on arresting my hostess for not reporting to them that she had let out her room for two days and concealing the fact that she had made some money (apparently somebody should have received a cut).

Meanwhile, Den Green had taken the rubber petrol tank in the trailer down the Adriatic coast to Kotor where they were to have the tank filled, but when they opened the trailer they found that the felt wrapping - which was supposed to have kept the tank from chafing - had disintegrated, and the tank had developed a leak. The cars would be relying on Den's fuel, on the other side of the ferry at Perast, so he found and purchased two fifty-gallon drums which he loaded into the trailer in place of the now useless tank. It was just as well that he did for BMC was the only team to have this service, and Den had to turn a deaf ear to the pleas of other competitors who would have to deviate from the

57

route and spend time obtaining fuel from a petrol station.

Healey held to ransom

While he was in Kotor, Den was supposed to collect an Austin-Healey that had been severely damaged in a collision with a taxi while recceing for the event. The car was in a garage where it had been made roadworthy. Den had been supplied with extra cash to pay for the repair, but when he asked for the bill, was told to wait while the garage boss made a phone call. The phone call brought two policemen in a car - into which Den was bundled and driven off to the police station. Once there he was put into an interrogation room with arc lights shining in his eyes. Through an interpreter Den was asked what he was going to do about compensating the taxi driver, and they mentioned a sum that was far beyond all the money that Den had with him. The police were under the impression that Den was desperate to take the Healey home with him, and spent all day trying to wring some money out of him, until at last Den told them that he would leave the car and report to his boss later. Then they let him go.

When Den reported his experience to Stuart Turner they decided the damaged car wasn't worth the hassle. But Belgian driver, Julian Verneave, was told that if he could get the car out of Yugoslavia when he was travelling through on holiday, he could keep it. Julian decided that to get his hands on even a damaged works Austin-Healey, it was worth the sum demanded and he duly collected it.

Paddy Hopkirk and Henry Liddon on one of the better Yugoslav roads.
(Courtesy Den Green)

This was the last time that the rally was held on the open road. It had first been run in 1927 as the Liege-Biarritz-Liege, the following three years it was Liege-Madrid-Liege before becoming established as the Liege-Rome-Liege until 1961, when it became Liege-Sofia-Liege for the event's final four years.

Servicing by boat, train and plane

1965 started off with another Monte Carlo Rally win, and this time Timo Makinnen and Paul Easter took the honours.

The weather, which could vary according to the starting point, had a profound affect on the Monte. Sometimes all of the competitors from one start would fail to make it to Monaco, while those from another would have an easier time of it. This year, Stuart had decided to spread the team throughout Europe, using starts as far apart as Minsk, Stockholm, Paris and Athens.

Two cars were starting from Athens - Geoff Mabbs in his privately-owned Mini, and Rauno Aaltonen with Tony Ambrose in the works Mini Cooper S. This was to be my starting point. Geoff and I drove the two cars down to Marseilles, where they were loaded on to the ferry. Geoff stayed in France to carry out further recceing, flying to the start later. I boarded the ferry bound for Athens. For many of the other passengers on the boat this was a Mediterranean cruise which would take them to Egypt, and I enjoyed two days of leisure, for much of the time in the company of a musical couple. The wife was a singer and her husband was the drummer in a band that played on the end of Southend pier in the summer season.

I was met at Piraeus, the sea-port of Athens, by two employ-ees from Doucas Brothers, the Athens BMC dealer. I carried out the usual pre-rally work in what was, by now, the familiar Doucas workshop.

After the rally start I was scheduled to carry out a service many miles away in Zagreb, Yugoslavia. My transport was the trans-continental railway from Athens to Belgrade. And so once more I set off with my bag of tools and my suitcase, this time to occupy a two-berth sleeping compartment by my-self (with my James Bond book to lend atmosphere to the jour-ney). I took a taxi from the sta-tion at Belgrade to the airport, from where I was due to make the next leg of the trip to Zagreb.

The airport was shrouded in fog and all flights were delayed. I had the phone number of Ral-ly Control in Zagreb and was able to contact the rally officials there, who gave me the informa-tion that the works Mini was on the start-line, ready to go. There was no other course open to me but to continue on my way to Monaco, and this I did by way of flights to Zurich, Paris and Nice. My abiding memory of this epic journey is of the meals that the different airlines gave me at the start of each flight.

Rauno made it through to France but soon after started to experience a bad misfire. The cause of the misfire proved diffi-cult to trace and, by the time it was found and rectified, Rauno was out of time.

The night after arriving at Monte Carlo the cars and crews had to face 400 kilometres of mountain roads at speeds that would turn every service halt into a pit-stop. This was a far cry from the first event I rode with Marcus Chambers, more as a spectator than a mechanic. The mountain roads were lined with rally enthusiasts in party mood, some of them having drunk rath-er more wine than was wise.

Toward the end of the night near disaster struck Timo. His car stopped with what his co-driver, Paul Easter, quickly diagnosed as a broken contact breaker. This was one of the parts carried in the car's kit of spares and Timo quickly fitted the new set, with-out losing enough time to deny him his outright win. The next morning Nobby Hall was dis-patched to collect the car from the overnight Parc Ferme. The Mini refused to start. A check soon showed that there was no spark: the feed to the distributor was going straight to earth. Fur-ther investigation revealed that, in his haste, Timo had neglected to refit the insulating washer be-tween the contact breakers. How the car reached the end of the rally was an absolute miracle.

Mini Monte in Sweden
Amid all the celebrations after we returned to England, one of the rally cars was exhibited at the Racing Car Show and we were all given time off to visit the show. Whilst there with my wife I was called on the tannoy. There was to be televised 'Mini Monte' in Sweden and, for some reason, I was chosen to accom-pany Bill Price, each of us driv-ing a Mini through Europe to the event, and we were to leave the next day. I had been away from home for a fortnight and asked for permission to take my wife along. This was granted and we made our way to Denmark, then on the ferry to Sweden.

The event was held on a frozen lake, which gave rise to some exciting racing. The cold was kept out by a regular supply of 'Glug', a hot, spiced wine.

I had been amazed on my first visit to Sweden to see traffic using roads made by bulldozing a path through the snow on the lakes. These roads were a per-manent winter feature and were signposted where there were crossroads on the ice.

Mick Legg and Timo Makinen
Mick was an ex-apprentice who had fitted well into the Competi-tions Department. He had a touch of anarchy about him and a flair for getting into trouble. As an ex-apprentice he was well known to all of the factory employees.

When testing a rally car we would use the same route as the production testers and Mick would take great delight in blast-ing past the testers at full speed. On one occasion he happened

Works Rally Mechanic

Acropolis Rally, 1966: Aaltonen/Ambrose leaving the start. (Author photo)

to see his mother, who was cycling from Abingdon to the village of Marcham. Mick stopped and, with his Austin-Healey door open, chatted to her for a few minutes, little knowing of the consternation caused when the testers reported that 'that mad bugger Mick Legg has knocked an old lady off her bike.'

Mick was sent out to Finland to service the cars recceing for the 1000 Lakes Rally. He was the only mechanic with the cars, and the three drivers - Timo, Paddy and Rauno - each wanted him to ride with them to be on hand should he be needed. It was agreed that whoever got his car and picked up Mick first should take him with them that day. Next morning it was Timo knocking on Mick's door. When Mick asked where the others were, Timo told him that he had taken the carburettors off both their cars and it would be an hour or so before they would get them mobile again!

The night before the rally started Timo and Mick had a evening out, not getting back to the hotel until breakfast. Stuart Turner was understandably livid. 'You're sacked, Legg!' he shouted. Timo butted in 'If you sack him you will have to sack me as well.' 'You're damn right I will', replied Stuart. 'If you don't win this rally you're both out!' Fortunately, the result was Timo first, Rauno second and Paddy sixth. And so Mick survived to get into trouble another day.

Later on we lost Mick to the Nairobi BMC firm of Ben-bros. One of the partners, David Benzimra, came to England to take part in a rally and invited Mick to go to Kenya to work for him. Mick took up his offer and spent ten happy years causing mayhem among the female population of the African continent.

Nobby Hall lifting the considerable weight of the Austin-Healey on the 1965 Targa Florio. (Courtesy Den Green)

Works Rally Mechanic

A Greek tragedy

A great feature of the Acropolis Rally was the boat trip there. In 1966 we drove down through Italy to catch the boat at Brindisi. A small problem with the Italian Customs was encountered before we were allowed on board. The trouble had originated at the border when we entered Italy, where tourists were given discount petrol coupons. The number that one was allowed to use depended on the miles trav-

Cruising to Greece: Tommy Wellman, Bob Whittington, the author, Roy Brown, Dudley Pike and Pete Bartram. (Author photo)

Acropolis Rally 1966: the author and Roy Brown servicing the Aaltonen/Ambrose car. (Courtesy Den Green)

and, although we never used the coupons, we always made a point of having them because we knew that we would be asked for them later.

The Customs official at the border had refused to give us any coupons because, he said, we were not tourists but were travelling through the country commercially. When we tried to get out at the other end, the Customs official there insisted that, as we were in private cars, we were

elled in the country. If too many had been used then the difference between the discount cost and the normal cost had to be paid. The unused coupons had to be handed in when leaving the country. We were aware of this,

Donald and Erle Morley in the 1964 RAC Rally. (Courtesy BMC)

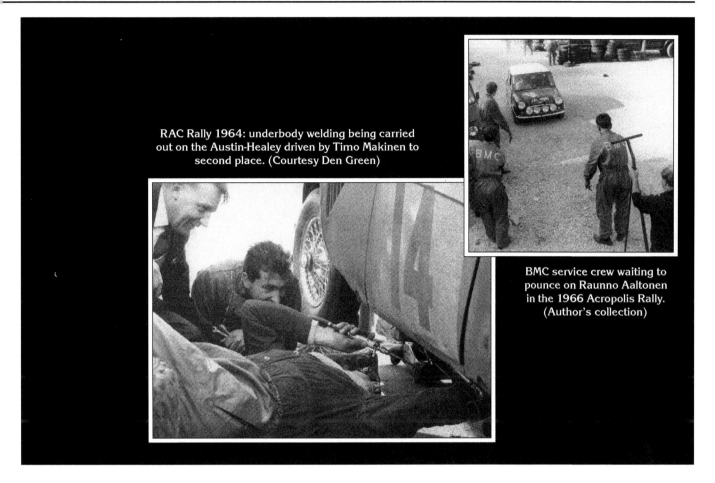

RAC Rally 1964: underbody welding being carried out on the Austin-Healey driven by Timo Makinen to second place. (Courtesy Den Green)

BMC service crew waiting to pounce on Raunno Aaltonen in the 1966 Acropolis Rally. (Author's collection)

tourists and must have been given coupons. The dockers were waiting to load the cars on to the boat and we were still arguing, until in desperation I asked how much we owed, whereupon I was invited into the office, given a glass of wine and the matter was settled in minutes.

The mechanics on the trip were, firstly, our foreman, Tommy Wellman. Tom had been well taught by Marcus in the gourmet arts and his capacity for pints of beer was matched only by Den Green, his deputy. Then there was Bob Whittington, who was always immaculately turned out:

when sunbathing he would religiously turn a different part of his anatomy to the sun at regular intervals to get an even tan. Then Roy Brown - a lay preacher in his spare time, who came in for a fair amount of ribbing (which he bore with good grace, along with his nickname 'the Vicar'). Dudley Pike was an ex-MG apprentice with what romantic writers would describe as 'rugged good looks' and was never short of girlfriends. Pete Bartram made up the party. He was eccentric: if something was not going right when he was working on a car, he would attempt to rectify it by

swearing. It didn't pay to take the mickey out of him in this mood, for he was prone to throwing his tools at you, but when the moon was in the right quarter he was an excellent mechanic.

I was sharing a barge with Roy Brown on this trip. We were due to service on a road junction. The control was about twenty metres down the minor road. Traffic on the major road made parking there very dangerous; consequently, servicing was being carried out perilously close to the control - in fact, one manufacturer's team had arrived early and occupied a parking space

Works Rally Mechanic

84-hour Marathon 1967: drivers Alec Poole, Roger Enever and Clive Baker failed to finish with LRX 830E. (Courtesy Mick Hogan)

Far right, inset: 84-hour Marathon 1967: mechanics on pit counter, Tommy Wellman, Peter Bartram and Mick Hogan. (Courtesy Mick Hogan)

Main pic: 84-hour Marathon 1967: cars ready for transporting to Spa. GRX 5D, driven by Tony Fall, Julien Vernaeve and Andrew Hedges, came second overall and first in class. (Courtesy Mick Hogan)

that was actually within the limits of the control.

It was early in the morning. The heavy, long-range driving lamps were mounted on a quick-release bar. All Paddy wanted was that the lamps and bar removed be carried by us to reduce his weight during the day. A man, whom I assumed was a spectator, told me that we were servicing within a control and the driver could be penalised. I pointed out that we were not alone in this as the lack of space left no alternative. This man was, in fact, part of the control team of officials, and when Paddy drove to the control table he was told that he would be penalised. I protested and the officials agreed not to take the matter any further, but said that if any other teams brought the matter up this decision would have to be reconsidered.

Paddy went on to become the winner. However, the driver for a rival team had been penalised for allegedly taking a short cut which deviated from the rally route and, it was very heavily rumoured, the manager of this team heard about the affair with Paddy and demanded that, if his driver was penalised, he would lodge an official protest against Paddy. The organisers decided that as the matter had now been officially brought to their notice, it would have to be dealt with, and our misdemeanour, coupled with the fact that Paddy had booked in early at a control, meant a penalty that deprived him of his win and dropped him

down to third place.

I was not popular, especially with Dudley, who had built Paddy's car and had started celebrating his first win.

Graham Hill - rally driver

The RAC Rally was not a popular event with the mechanics. It was cold, muddy and went on for four days with one night stop, which didn't give us much respite. Servicing was carried out at the last control before the cars were locked away for the night, and it could be gone ten o'clock before the last car was safely put to bed; then we had to get cleaned up before finding something to eat and getting to bed. After a too-short sleep, it was up again to be at the first stop before the first car arrived.

Where it was logical for one team to service applied equally to other teams, and we had to get to the rendezvous early to secure the best spot. This might be on a garage forecourt, after discussion with the proprietor, or maybe a pub car park (a very popular choice, this one!).

There were official Service Areas, which were very crowded with all the 150 or so competitors coming and going, and it was important to get a good pitch in these. If you were too far in it would be difficult and time-consuming for your car to reach you, and the same applied when it tried to get out. A notorious spot was at Plashetts in the heart of the Kielder forest. This was the coldest, muckiest part

of the whole rally, and trying to find hard standing to jack up a car up was a nightmare.

The Sun newspaper sponsored the 1966 rally, and two of the leading Grand Prix drivers, Jim Clark and Graham Hill, were driving for the two leading rally teams, Ford and BMC respectively. This generated lots of extra publicity.

Hill was at the wheel of a Mini Cooper S and wasn't enjoying it one little bit. He couldn't believe how fast the cars were driven through the forests and was convinced that all rally drivers were mad. No doubt he was greatly relieved when his transmission failed, causing him to retire.

Half our entry of eight cars failed to finish, but of the four that did, Harry Kallstrom, Rauno Aaltonen and Tony Fall were second, fourth and fifth respectively. I was particularly pleased, as I'd prepared Harry Kallstrom's car.

Party time

The BMC Competitions Department Christmas parties were legendary and were organised by Bill Price and the mechanics. A local hotel would be booked for the occasion, and all of the drivers were invited. After the meal the fun and games would begin. A comic song was always composed, highlighting any amusing or embarrassing events that had occurred during the year, and presentations would be made on the same theme. Peter Bartram had had the misfortune on one rally to set fire to a car when it

Bete Bartram wearing his firefighter hat. (Author photo)

very silly, but in the party mood it all seemed very funny.

The butt of many of our jokes was often Henry Liddon. Henry was a big man and could always be guaranteed to eat everything in sight when he arrived at a service point. This was ref erred to in the carol sung to the tune of Hark the Herald Angels Sing:

Robin Vokins, Tony Ambrose, Andrew Hedges, Stuart Turner and the author singing 'Hark! The Mini gearbox sings'. (Author photo)

was on its side with petrol leaking from the filler cap, and he was awarded a fireman's helmet and extinguisher. At an after-rally party I had turned a go-cart over and for this I got a model go-cart and a crash helmet. All

Hark the Mini gearbox sings
Rauno Aaltonen it brings
To our humble service point
Contemplating his Birfield joint
Backwards round the corner skating
Left foot braking demonstrating
Keep the ham and chicken hidden
From his navigator Liddon.
Don't let him know how much

Competitions Department annual dinner, 1967. A 'send-up' of a Rally Service Point. (Author photo)

Works Rally Mechanic

Competitions Department staff and wives at the annual dinner in 1965. (Author photo)

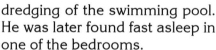

Competitions Department Stores Foreman 'Santa' Neville Challis hands a present to the author's wife, Iris. (Author photo)

we've got
Or he will scoff the bloody lot!

Stuart Turner threw himself into the spirit of the party, helping with scripts and suggesting ideas. It was Stuart who suggested a sketch depicting a farcical service point, with Stuart, Andrew Hedges and Tony Ambrose taking the parts of mechanics. I played Peter Browning, with an alarm clock because he was a race timekeeper. Robin Vokins was dressed as Stuart's secretary. Doug Watts was Timo and Roy Brown his navigator. What Mick Legg was doing on the tricycle I've no idea. The cardboard Mini had been made in the workshop (see picture previous page). Summertime brought the barbecue. Sandy Lawson had joined the department as Peter Browning's secretary. Her family owned a farm between Abingdon and Oxford and they kindly allowed us to take over the farm for

the night of the barbecue. The farmyard was conveniently set out with a large barn forming one side of a quadrangle round a cobbled yard. Everybody paid a sum of money to cover the expenses. There was a super-abundance of food, with Tommy Wellman cooking the steaks, and the drink was on a help-yourself basis.

On this first barbecue we had dramas galore. It poured with rain, making the cobbled yard treacherous. Hamish Cardno, a motoring journalist, slipped and cracked his head on the cobbles. An ambulance was called, and when it arrived it came screaming into the farmyard tearing down all the lights that we had strung across. John Milne, Bill Shepherd and John Williamson had all come down from Scotland for the event. John Williamson went missing, causing a panic-stricken

dredging of the swimming pool. He was later found fast asleep in one of the bedrooms.

The MG Company accountant, Norman Higgins, the man who had to make sense of our claims for expenses after every rally, disappeared after telling somebody that he was going to walk home. His wife, Jean, was not too pleased to hear this and got in her car to look for him. Soon, she telephoned to say that she had got all the way home without seeing Norman. A small search party comprised of those still capable of driving set out to look for him. He was found almost in Oxford, after having turned left instead of right out of the farm gate.

The DJ fell asleep about two o'clock, but nobody seemed to notice, and, come the dawn, it

Tommy Wellman (right) and Brian Moylan at a Competitions Department barbeque. (Author photo)

toon and, seeing that I was interested, invited me to join them. I couldn't get used to their way of dealing. Unlike us they dealt from the bottom of the pack. Every time that I had the deal I would start from the top, which caused great uproar especially as I was doing most of the winning. But it was all good-natured and I was spending all my winnings on drinks for everybody so it was a very convivial evening.

Once the rally had started we were to drive through Greece to Belgrade in Yugoslavia, where there was a main control, and carry out a service. Our instruc-

Paul Easter, Timo Makinen's co-driver, watching preparation of their 1966 Monte Carlo Rally car, which, after being provisionally announced the winner, was later disqualified on a technicality. (Courtesy Paul Easter)

was unanimously voted an excellent night and one that should be repeated next year. It was, and for every year after for as long as the Department existed.

Monte skulduggery

In 1966 we had every right to say 'We wuz robbed!' The three Minis had finished in the first three places in the Monte Carlo Rally and all three were disqualified. The reason put forward was that their lights did not conform to regulations.

I had been sent to the Athens start again, this time in an Austin A110 Westminster barge with Roy Brown. Our outward bound travel itinerary had us taking off from Southend airport, complete with Westminster, to land at Geneva; from there we drove to Milan. The next leg of our journey was by train to Rome. Then a day's drive across the foot of It-

aly brought us to Brindisi, where we boarded the ferry to Greece.

This time the cruise was in a boat transporting trucks. The truck drivers had a card school going; they were playing pon-

tions were that we should not run with the rally, but this was very difficult as we were going the same way and, having set off just after the rally start, had to reach Belgrade before the cars.

When we crossed into Yugoslavia the weather turned very nasty, with deep snow and icy roads. Fortunately, the road we had to follow was the main road right up from the border through Belgrade and beyond to Italy. So this road, at least, was kept moderately clear. Petrol stations were a rare sight and at one we witnessed an amazing sight. A line of trucks had been parked up for the night at the roadside and, as we approached, we could see a line of fire. To beat the cold and save their engines from freezing, each driver had lit a bonfire under his truck. We drove quickly past.

Stuart had anticipated that we would be pretty shattered after sharing the driving between us for the long journey to Belgrade, so he arranged for Johnny Organ to be flown down to meet us to assist with our servicing and share the driving on the long trek to Monte Carlo. The appalling weather meant that a lot of flights were delayed, and when John arrived he had missed a night's sleep and was more exhausted than we were. He was too late to join in what little work we had to do and we set off straight away with our relief driver fast asleep on the rear seat!

Before the final test on the mountain circuit, BMC and Ford were informed that the eligibility of their lighting would be a matter for jurisdiction. Mick Legg and John Smith, the Lucas technician, were, somewhat ill-advisedly, despatched to change the headlight systems back to standard, but the press were following every move made by the Minis and only one of the cars had its lights changed because, the next day, photographs of Mick carrying out the change appeared in the French newspapers, confirming, the press alleged, cheating by the BMC team.

Basically, the trouble with the lights was due to the fact that this was in the early days of quartz iodine bulbs and they had not been developed with twin filaments; consequently, to dip meant switching down to auxiliary driving lamps. The rules on lighting were that they had to conform with the laws of the country of origin of the car, and there was no problem with that. It seemed that the rally authorities had simply decided that it was time a French car won.

Obviously unsure that a disqualification on the light 'infringement' would hold water, the scrutineers went to exceptional lengths to find other discrepancies. In this endeavour they failed, to the obvious embarrassment of the French mechanics who were carrying out the inspection. We probably didn't help our cause when we offered to measure the door handles and count the wheelnuts ... The outcome was that the organisers stuck to their guns in disqualifying the three Minis and the Ford. Fifth place went to a French car whose driver had gone home and refused to come back to accept the winner's award.

Despite the fact that we were not an official winner, the cars were flown home, with all the personnel, in a Bristol freighter. But my rally was not over. Timo's car had reached the end of the rally in a state of complete exhaustion. Amongst other things the seals had gone in the brake calipers: the last part of the drive from Monte Carlo to Nice airport had been with the handbrake only.

As the 'winner' this was the car that would receive all the publicity and it was scheduled to be unloaded first as soon as we touched down at Heathrow. The calipers had to be changed inflight as BMC did not want the car to be pushed out of the plane with no brakes into the full glare of the television cameras. Den Green and myself were allocated the task, which we weren't very happy about because everybody else was drinking champagne. The car was in the nose of the freight compartment in readiness to be first off. There was a two-inch gap in the floor and Den said 'For God's sake don't drop a bolt or we'll kill some poor bugger.'

While the cars and the rest of the team were going home in style, Bill Price was assigned the job of driving the transporter - loaned to BMC by the Cooper Car Company - which was filled with all of the equipment and wheels, plus two trailers and the spares off the other service vehicles that were on the plane. Bill was accompanied by Stan Bradford, the Department panel beater, and Robin Vokins in an Austin A 110. On the way, at the road junction of the A44 with the N4, they were stopped by a *gendarme* and advised that the road was closed to heavy vehicles. The *gendarme* suggested an alternative route which would add just 28 kilometres to the journey.

Very soon the pair passed a road sign with *Barriere de Degel* on it, which didn't mean much to them, so they carried on. Just outside of Laon they were again stopped, this time by a motorcycle patrol from the *Gendarmerie Nationale*, who indicated that they could not proceed because of the *barriere de degel* and that they should follow him to a lay-by where the transporter was to be parked. Bill was told to use the Westminster to follow on down to the police station. At the police station the duty sergeant explained that a *barriere de degel* was in force and meant that all vehicles over six tonnes were barred from the roads, which were susceptible to damage during the thaw following on many days of hard frosts. They would require a permit which would only be supplied on production of a weight ticket which was to be obtained from the *Ponts and Chausses* office. So again following their motorcycle escort they took the transporter to a weigh bridge. When the ticket came out

of the machine the *gendarme* burst out laughing! The transporter weighed just over fourteen tonnes and there was no question of a permit being issued. Bill was informed that the vehicle could not be moved until the *barriere* was lifted. He arranged for the local MG dealer to park it in one of their buildings and they went home in the Austin Al 10.

A week later BMC was informed that the restrictions were lifted, and Bill went back by plane and train to drive the transporter home, leaving a set of studded MG tyres with the MG dealer as a token of thanks.

Later on in the year the Abingdon police presented a letter from Interpol asking if W R Price and S Turner accepted responsibility for the offence of ignoring the road sign, which they did, and in October the case was heard in France, attended by French BMC lawyers. A fine of 200 francs was imposed, with costs of 92.35 francs.

Another Monte win

A tense atmosphere prevailed the following year when it was Monte time again. Some petulant voices were raised advocating a boycott. But pride and a feeling of 'we'll show 'em' won the day (to say nothing of the favourable publicity that the Mini had got out of the affair).

I had been preparing a recce car when building of the Monte Minis was started. When this was finished, Doug Watts asked me to work with Roy Brown on Rauno

Aaltonen's car.

Two men working on a Mini get in each other's way. The sequence of building was body and subframes first, engine last. Roy had started on the bodywork so it was convenient for me to build the engine on my own bench.

Engine building wasn't just a matter of getting all the parts out of the stores and bolting them together. First, the cylinder block had to be bored out to increase engine capacity, then the top face machined to 0.010 inches above the level of the piston crown. The specially balanced crankshaft was fitted next, but, before the pistons and connecting rods were fitted, the rods had to be carefully selected. A group one engine could not have any balancing or lightening carried out, so, in order to achieve a balanced engine, Neville Challiss' store carried a great many con-rods from which we were able to select a matching set (to match they had to be as light as possible, the small ends all had to weigh the same, as did all the big ends, and the overall weight of all the rods had to be the same, too. It was therefore an advantage to be the first to have pick of the stock. On a modified group two engine, weight balance was achieved by filing and polishing the rods.

Fitting the rods was not the straightforward job that it is normally. Before tightening the big end bolts, their length had to be measured with a micrometer; they were then tightened to the appropriate torque setting and

Works Rally Mechanic

Paddy's ex-1965 Monte car used for development testing. (Courtesy Den Green)

Bob Whittington balancing a set of conrods for the engine of the Mini Cooper 'S', which was second in the Monte Carlo Rally 1966 before disqualification. (Courtesy Den Green)

Monte Carlo: the damage caused to Timo's car by the mysterious rock.
(Courtesy Peter Browning)

Timing sprockets were made to align
without resorting to shims.
(Author photo)

Monte Carlo 1967. L-r: Roy Brown, Rauno Aaltonen, Henry Liddon and the author congratulating each other on the 'Monte' victory. (Courtesy Peter Browning)

tion. We were able to take advantage of this to make the volume of each of the four combustion spaces the same, by judicious grinding. Having ground out some of the metal from the combustion spaces, it was necessary to measure the combustion chamber's capacity. From this we could calculate the amount that should be machined off the cylinder head in order to get the required compression ratio.

After a trial fitting of the timing gear, we measured the amount to be machined off the camshaft gear in order to line up the gears without the use of shims.

the amount of stretch measured. This should be 0.003 inches and, if more, meant it was possibly a faulty bolt that had stretched beyond its elastic limit; if less, it was probably too hard. Either way, it had to be taken out and thrown away. This obsession with the big end bolts stemmed from the 1965 event when the car prepared for Raymond Baxter broke a big end bolt when en route to the start at Minsk.

Some cleaning of the casting in the combustion spaces of the cylinder head was permissible and was indeed carried out in the final stages of standard produc-

Monte Carlo 1967: Rauno Aaltonen and Henry Liddon with the winning car, which was quickly flown back to England. (Courtesy Peter Browning)

Monte Carlo 1967: celebrations in the MG factory. (Courtesy Peter Browning)

The gearbox was machined flat to accept a 'Moke' sump shield which was screwed straight to the bottom of the gearbox case. The build of the gearbox with straight-cut gears was straightforward, except for the selector locating screws with their locknuts - these had been known to come loose. The gearbox has to be completely dismantled to get at these so they were wire-locked in place to pre-empt any problem.

Four Minis qualified to start the Mountain Circuit. Again, the atmosphere when a car arrived for service was more akin to a racing pitstop than that of a rally.

With four Minis in the hunt all the ancillary teams - Ferodo, Lucas, Castrol, etc. - were anxious to help in every way they could, because the publicity generated by a Monte win was highly beneficial to them. Consequently, there could be too many well-meaning hands giving assistance. Time's car came in: amongst other things he wanted his rear brake linings checked. The Ferodo technician and I jacked up the back end of the car and were able to report that his linings were in good enough condition to finish the rally. We refitted the wheels, pinching the nuts up ready for final tightening when they were on the ground (the jack was in use at the front). Another car came in. I went to start work on the new arrival when I saw Time's Mini go screaming off into the night. I sought out the Ferodo rep and asked if he had given the wheel nuts a final tighten; he hadn't. We convinced ourselves that we had probably tightened them sufficiently anyway.

Ten minutes later we heard the news: Time had crashed. My heart gave a lurch - of course, I thought a wheel must have come off, but this was not the case. On a stretch of road unsuitable for spectators (due to the steep mountain side) and with no witnesses, a large rock rolled into the road into Time's path. The vulnerable front end of the Mini smashed into the rock, demolishing the oil cooler and distributor.

Rumours abounded among the rally fraternity: somebody had again sabotaged BMC's chances. But this was to be

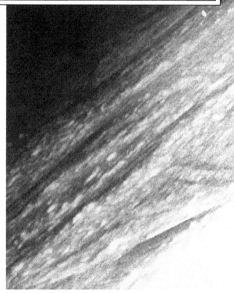

the only casualty and to everybody's delight, especially Roy's and mine, Rauno won in the car that we had built.

After the 1966 debacle BMC was making the most of this vindication. Once more the car was flown home for what was now its customary appearance on the revolving stage of the London Palladium. There was one small hitch. The car was in the wings waiting to driven on but it wouldn't start. Roy and I were summoned backstage. The clutch plate, punished by the clutchless gearchange technique, had shed dust all around the inside of the clutch/flywheel housing, clogging the starter engaging cog. The car's tool kit was still in the boot, so we were able to remove the starter swiftly and clean the bendix in time for the grand finale of the show.

All the cars and the team had been flown home. We were met by our wives at Heathrow and, after the Palladium show, were all taken for a meal and party at the Flanagan's 'sing-along' restaurant in Kensington, where a pia-

Peter Browning, Les Needham, Robin Vokins, Don Hayter, Bob Whittington, Tommy Wellman and Johnny Evans with the record-breaking Austin 1800 at Monza. (Courtesy Den Green)

Finnish rally ace, Rauno Aaltonen, races a Finnish skiing ace downhill on snow; the result was a tie.
(Courtesy Den Green)

nist played all the old music hall songs. The words were printed on the table napkins and the customers all joined in. On the plane home, Dan Dailey, whose company had made all the travel arrangements, asked us all to compose words to the tunes of the old favourites. He had these typed up and photocopied so that at Flanagan's we could sing our own words when the pianist struck up the appropriate music. One typical offering showed

how we had accepted the Timo trauma with a song to the tune of Clementine. The last verse went as follows:

No-one knows who shoved the rock down
But Timo said he thought
It must have been the 'little' people
Giving Rauno their support!

Strange happenings at Monza

In 1967, BMC - with backing from Castrol, who provided 60 per cent of the cost - instructed the Competitions Department to look at international records set by cars in the 1500-2000cc range, with a view to attempting - in an Austin 1800 - to break some of the records held by that class of car. The records in question were for endurance over four, five, six and seven days, plus 15000 miles, 20000 and 25000 kilometres. The attempt was to be made at the Monza Circuit in Italy in a car that had been specially modified for the task.

The time dragged by for the support team while the car was driven for twenty-four hours a day, with brief periods of activity when it came in for refuelling and driver changes. The drivers, too, were affected by the boredom of circulating the banked Autodrome, albeit at 100-plus mph. And so a number of tricks were devised to play on the drivers to help keep them alert.

First they were asked by a radio link to solve some mental arithmetic questions. Rauno Aaltenonen seemed to be the most proficient at this, until it was learned that he had worked them out with his finger in the dust on the dashboard. Tommy Wellman had designed a system with an electric pump that topped up the engine oil while on the move; unfortunately, Alec Poole hung the radio mike on the switch, causing the engine to overfill and black smoke to come billowing out of the exhaust.

Night time was when the drivers started seeing apparitions, and one of these was Ron Stacey of Castrol, who wrapped himself in a sheet and hung from the signalling gantry like a ghost. In the early mornings a mist would rise from the ground. Roger Enever, son of the head of the MG Design Office and an accomplished racing driver, was coming to the end of a stint at the wheel after a long night. He was amazed to see a naked nymph arise from the mist at the side of the track, but when asked if there were any incidents to report, was reluctant to admit that he had been seeing prancing fairies. It was not until later that he was told that the 'fairy' was indeed Alec Poole (who had hidden naked in the bushes until he heard the car approaching).

An elaborate joke was played on one of the drivers. When he drove into the pit area it was normally full of the clutter - chairs, umbrellas, spare parts, refuelling equipment, etc - of seven days of occupation by about twenty people. This was all whisked out of sight, so that the bemused driver passed by an utterly deserted and empty pit, only to find it all back to normal on his next lap. He was very quiet for the rest of his stint and afterwards admitted that he thought he had imagined the whole thing.

The week ground on towards its end and the average speed in that time was 104mph. RAC timekeepers were on hand to verify the speeds, and it was decided that the car was going too fast on the assumption that the motoring public would not possibly believe that a car widely owned, with a usual top speed of 90mph, could achieve the high speeds claimed for it.

The records had been soundly broken, all at an average speed of 92-93mph, affording BMC and Castrol a great deal of publicity material. And the lads had good cause to celebrate.

Twin Weber carburettors

The 1968 Monte Carlo Rally - Stuart's last as team chief - saw us falling foul of the Monte scrutineers again. Racing Minis had long been using a single twin-choke Weber carburettor, with a notable gain in horsepower compared to the twin SUs which were the standard fitment. The regulations for that year's rally allowed a change of carburettor so long as there were the same number of them and they were mounted on the standard inlet manifold.

The Weber had a float chamber that fed both chokes of the carburettor. One choke and the

mounting flange were machined off and a flange suitable for the inlet manifold was welded on and a new 'single' choke carburettor was the result. It was, of course, necessary to cannibal-

ise two Webers to make a pair of carburettors that were within the rules. Weber was made aware of the exercise to ensure there wasn't a problem over patents.

The Monte officials were not overjoyed at having to admit that the modification was legal, but as the top placed Mini was third and not first, the result was allowed to stand.

Timo had been the only casualty on the run in to Monaco and three Minis were qualified to start the mountain circuit.

There were fourteen different patterns of tyres available with differing lengths of studs and spikes to suit all road conditions. One type had long thin pointed spikes for use in deep soft snow which could only be handled wearing stout gloves, and we called them 'Hedgehog tyres.' There were so many tyres that a Dunlop truck was used to transport them to strategic points around the circuit.

Gone to the pictures

Stuart Turner's briefing told us that the first service point in the evening would be extremely congested, with service crews and spectators all trying to find vantage points, and we were instructed to get into position very early.

Derek Plummer, another ex-apprentice, and myself got to this

Under the bonnet of a 1968, twin carburettor Monte Carlo rally car.

first point early in the afternoon and reserved a space for the Dunlop truck; then we had a meal and prepared ourselves for the long night ahead. Stuart arrived nearer the time when we could expect the first car and was astonished and angry to find that the Dunlop truck had not put in an appearance. Time went by, then half an hour before the first car was due, the truck arrived. Gordon Pettinger, the Dunlop tyre fitter and truck driver, had looked at his schedule and - seeing that he had all afternoon to waste - went to the pictures. Gordon was completely impervious to all criticism and discipline, and on Stuart's demand that he explain himself said in his wonderful broad Brummy accent

'Well ahm 'ere now ent I, wot am yow worried about?' There was no answer to that.

Farewell party for Stuart Turner on his leaving to join Castrol. (Author photo)

Wrong victim

A dramatic incident occurred towards the end of the Mountain Circuit. Some local revellers, spectating on the Turini hillclimb, attempted to enliven proceedings by depositing a large amount of snow on a fast corner that had been dry when the drivers had tackled it first time round. Unfortunately for them, the first car to come to grief on the unexpected surface was the French ace Gerard Larrousse, who was leading the rally and was set to notch up the first major win for his team, Renault Alpine. The culprits were quickly identified and only the intervention of the *gendarmerie* averted a lynching.

Larrousse's retirement left the way open for Vic Elford, one-time BMC navigator, driving a Porsche to take first place in the rally

Farewell Stuart

This was to be Stuart Turner's last Monte in charge of BMC's Competitions Department before he left to take up an appointment with Castro), but he was soon after snapped up by Ford to take over its Motor Sport programme.

We had a collection and bought a silver salver, on which each of our signatures was engraved, as a leaving present. As a further memento of the Department's achievements under his leadership, a unique desk set was made with Corgi models of the four Monte Carlo Rally-winning Minis (1966 was still considered a win in our eyes).

Stuart's departure was the excuse for another party, which was held in the MG Social Club ballroom. All round the walls we hung spoof Castro) posters, based on genuine examples. Stuart's photograph was substituted for 'The Laughing Cavalier' on their poster with the title 'The Masterpiece in Oils'.

Tour de France 1964: the Aaltonen/Ambrose Mini Cooper 'S' being serviced by Tommy Eales. (Courtesy Paul Easter)

Timo recceing for short-cuts. (Courtesy Paul Easter)

Colour Gallery

Liege 1964: Paul Easter on the Obrovac road after dropping off the author. (Courtesy Paul Easter)

Liege 1964: Den Green with the petrol trailer. (Courtesy Paul Easter)

Typical Yugoslav traffic jam! (Courtesy Paul Easter)

Colour Gallery

Liege 1964: BMC mechanics' route conference. (Courtesy Paul Easter)

Timo Makinen with Paul Easter on the 1965 Scottish Rally. (Courtesy Paul Easter)

Mechanics checking the results of sump shield testing. (Courtesy Paul Easter)

Minis awaiting embarkation after the 1967 Acropolis Rally. Note the 'A' bracket on LBL 606D, which enabled a car to be towed without a driver.

This slightly fuzzy original photograph gives us a rare glimpse of Minis in the 'Comps' workshop. Number 99 is Timo's 1967 Acropolis car, and JMO 969D is Paddy's Circuit of Ireland car. (Courtesy Paul Easter)

Tony Fall's 1967 Geneva rally winner having its bottom inspected. (Courtesy Paul Easter)

A rather dark, but rare, photograph of the 1800s unloading on arrival at Nairobi Airport. (Courtesy Mick Hogan)

Colour Gallery

Tommy Eales, Bob Whittington, Gerald Wiffen, the author, Doug Watts and Dudley Pike with their Safari guide.
(Author photo)

Left to right: Doug Watts, Dudley Pike, Gerald Wiffen and Mick Hogan on the road to Mombasa.
(Courtesy Mick Hogan)

Parting of the ways: camels to the left, cars to the right. Paul Easter, with the car he shared with Rauno Aaltonen and Henry Liddon. The winch on the front is holding the suspension together. (Courtesy Paul Easter)

A typical East African village. (Author photo)

Mick Legg with his admiring audience.
(Courtesy Mick Legg)

A time control on the 1968 London-Sydney Marathon and all's well with the car driven by Rauno Aaltonen. (Courtesy Paul Easter)

Colour Gallery

BMC service in outback Australia.

The Peter Jopp, Willy Cave and Mark Kahn Special Tuning-prepared Austin 1800 being services by Doug Watts on the 1970 World Cup Rally.

Colour Gallery

London-Sydney Marathon 1968: the service point which was set up on the edge of the desert by the Tehran BMC dealer Ravand & Co. Ltd. (Courtesy Paul Easter)

Tommy Eales (left) joins Bob Whittington (right) and Bill Price in the 'sweep' car to Bombay. Desert recce car on the left. (Author photo)

Peter Browning

Of the three BMC Competitions Managers I think I probably drew the short straw, arriving on the scene when the Mini was starting to lose its advantage to the European opposition, when the big Healey was out of production and the MGB was really only suitable for chasing class awards on the race circuits. Then came the British Leyland merger, and all of our budgets went on the new generation of Marathons with the 1800s on London-to-Sydney and 2.5 Triumphs on the London-to-Mexico. Runners-up on both events was too much for Lord Stokes to bear and, of course, then came the sad closure of the best rally team in the world. It was typical of the team, particularly those who had achieved so much through the 'winning' years, that they never lost their enthusiasm and dedication for the job - indeed, they became more determined for continued success. Brian Moylan, one of the best in the business, was a typical example, and his behind-the-scenes stories will jog a few memo-

Peter Browning

ries of incidents that most of us had forgotten or are reluctant to agree were actually true!

Peter Browning

Safari debacle

The Coronation Safari, to give the event its original title, was first run as part of East Africa's celebrations to commemorate Queen Elizabeth II's coronation in 1953.

It achieved international status in 1957 and, in 1960, the name was changed to the East African Safari. Local knowledge was all-important; in all the years that the event had run up till then there had never been a non-African winner.

Africa was an emerging nation and a fertile market for rugged, reliable cars. It seemed to BMC that the Austin 1800 could be a prime contender for this event and, to this end, an intensive development programme was carried out.

Cliff Humphries looked after this side of the business, when he was not building rally car engines. He and Eddie Burnell (Eddie was our test driver) spent many weary days driving round the punishing Army tank-testing circuit at Bagshot. Eddie had been recruited from the Service Department and, although small in stature, was as quick as any of the rally drivers when testing on a circuit or a forest track. His lack of height was apparent when

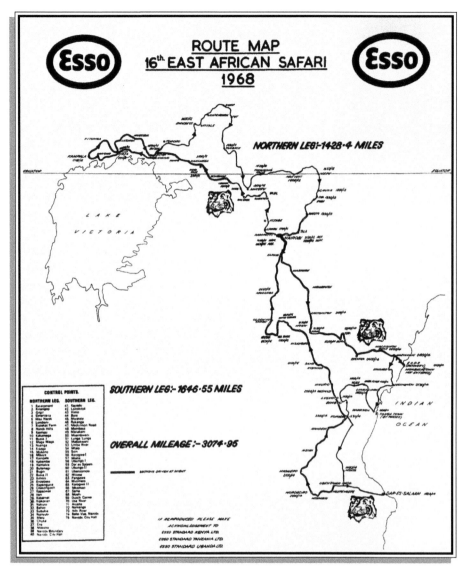

ROUTE MAP
16th EAST AFRICAN SAFARI
1968

NORTHERN LEG:-1428·4 MILES

SOUTHERN LEG:- 1646·55 MILES

OVERALL MILEAGE:- 3074·95

was the first time that I had seen armed police guarding the airport - alas, an all too familiar scene nowadays.

Arriving at Nairobi, we were taken first to the garage of Benbros Motors, which was to be rally headquarters for our stay. Here, we met the workshop staff, a mixture of Asians and Africans and a cheerful enough bunch. Then we were taken to our accommodation, a block of apartments on the outskirts of the city owned by the Nairobi Country Club. We were cared for here by a full-time staff, who looked after our rooms, did our laundry and cooked our meals, which we ate in a communal dining room.

The drivers started practising with the cars and very soon showed up some flaws that had not become apparent during the Bagshot testing. The front suspension hydrolastic units on the 1800 are in an aluminium housing and it was on this housing that the front Aeon rubbers were fitted. On the East African roads, where the wheels were off the ground more than they were on it, the constant leaping and landing was more severe than anything encountered at Bagshot, and the hard rubbers were transferring the shocks to the housing, causing them to break up. The whole suspension was geared to the use of the rubbers and attempts to change the setup were unsuccessful in the time at our disposal. So it was decided to run with it and fit new hous-

one saw him mounted on his regular form of transport, a Vincent HRD 1000cc motorcycle.

Every day the pair were able to break some part of the car's suspension, or a new fracture would appear in the bodywork. The constant crashing over the rocky terrain burst the overworked hydrolastic suspension units. Additional rubber buffers, called Aeon rubbers, were

mounted to help absorb the shock. These heavy, concertina-shaped rubbers were fitted between the frame and the suspension arms and had to be compressed to get them in place. This was not an easy task which proved even more difficult in the heat of the rally.

Our rally party flew out to Nairobi in a chartered Britannia, stopping to refuel at Bahrain. This

Typical flash flood encountered while recceing for the 1967 East African Safari. (Courtesy Mick Hogan)

throw out of the car as payment for this service.

To help smooth our way in this environment (which, to us, was totally new) we were provided with a local man who was well travelled in the country. This was a kind thought, but his presence was to prove an embarrassment when our way took us into Uganda. Our service point was in sight of Lake Victoria. It was here at one o'clock in the morning that we had our first experience of fitting a new suspension housing under pressure. It was a beautiful moonlit night when Timo Makinen arrived to have the job carried out. No sooner had we jacked up the car when, with no warning, the heavens opened and in minutes we were paddling up to our ankles in water and our tool boxes, handily by our sides, were filled to the brim. The job was done but Timo was destined not to go much further as he suffered a burst oil cooler. Our service completed, we wanted to waste no time getting back into Kenya and on to our next point.

Getting into Uganda had posed no problems, armed as we were with the right visas. Getting out again was not so easy. Tommy Eales and I had exit visas, thanks to Peter Browning's impeccable attention to detail. Our local guide, however, was not so equipped - and we were held up for two hours while many telephone calls were made to establish the *bona fide* of our passenger.

The cars of Rauno Aaltonen

ings at every opportunity.

The roads were of red murram dust that turned into a quagmire in the tropical rain storms that struck without warning. Dried river beds turned into raging torrents that had to be forded. Everything under the bonnet was waterproofed, a snorkel fitted to feed air to the carburettor and a pipe used on the exhaust to raise it above water level. The cars were frequently stuck in mud, and, in anticipation of this, handles were fitted to the front wings for the navigator to hold whilst standing on the special footplates on the front bumper and bouncing to gain traction. In isolated and apparently uninhabited spots, a car stuck in a ford would immediately be surrounded by native children who, by sheer force of numbers, manhandled it to the opposite bank. The navigators carried a stock of small change to

Tony Fall failed to make it back to Nairobi on time. (Courtesy Den Green)

Nairobi Railway Station, gateway to Mombasa. (Author photo)

The Elephant who had the last laugh on Mick Hogan. (Courtesy Mick Hogan)

and Tony Fall suffered from broken hydrolastic unit housings, and time taken to have them changed put them outside the permitted lateness at the halfway stop back in Nairobi.

Holiday time

So there we were, back in Nairobi with no cars left in the rally, and the Britannia to take us home not due for a week. But we were in Africa. Most of our previous travelling had been confined to Europe, so it was an experience that many of us would not have the chance to repeat, and Peter allowed us to take full advantage. We had a day's safari in Tsavo National Game Park, a visit to a Masai village (where we bought lots of carved wood souvenirs), then a trip down to the coast of Mombasa. Most of the party travelled down by train but Mick Hogan and Dudley Pike drove one of the service cars, and had a frightening experience with an elephant.

Elephants were not a rare sight, but were not as common as one might have expected. Driving through the heavily wooded countryside they suddenly saw a magnificent bull elephant in a clearing just off the road. A wonderful photo opportunity, they thought, but the animal turned its back on them in complete indifference. Not to be thwarted, the lads got out of the car and shouted to attract its attention. This proved to be only too successful . . . the elephant turned and, with a loud trum-

peting, lumbered towards them. Dudley leaped back into the driver's seat and had the car moving; Mick was only halfway in, clinging to the door, as they sped off. The elephant, meanwhile, had accomplished his objective and stood in the middle of the road watching their retreat. They swear he was laughing at them.

BL Competitions press officer, Alan Zafer, keen to get some action shots, had Peter driving towards him on a track in the Nairobi Game Park. The picture he was after caused him to

A Mombasa beach bungalow.
(Courtesy Mick Hogan)

retreat off the track, almost into the bush. After several dummy runs, Alan backed up to a thorn bush and could go no further. He got his shot and Peter drove across the grass to pick him up. As they drove past the bush to the track, there, on the other side of the bush - not ten feet away -was a lioness and cubs. Alan, understandably, broke into a sweat when he realised how close

to danger he had been.

Mombasa was a holiday paradise. We lodged in a hotel complex with bungalows on the beach, the Indian Ocean lapping the sands just yards away from our verandas. The local fishermen would bring ashore many strange fish, but their superstitions prevented them from allowing us to take any photographs.

Back in Nairobi all the cars and spares were loaded into the Britannia, which had had most of its seats removed to make room. The weight of everything must have taken the plane close to it's maximum capacity.

Nairobi airport is 3000ft above sea level, and the Britannia trundled along the runway until we neared its end. It seemed that the pilot lifted the nose of the plane more in hope than in any genuine anticipation that the airscrews would find enough purchase in the rare atmosphere to get us airborne. The wheels relinquished their hold on the ground and, slowly but surely, we gained height. We headed home, having been made dismally aware that Bagshot testing was no substitute for the real thing.

A desert crossing

The exciting event of 1968 was the London to Sydney Marathon. Austin 1800s were to be used again and, with the experience of the Safari to draw on, Bagshot testing was carried out to better effect. A recce of the route was undertaken by the BMC navigators and they reported back on the road conditions encountered.

This information greatly assisted the team carrying out development of the cars for the event. Nobby Hall had joined Cliff Humphries and Eddie Burnell to form this development team. Nobby was an excellent choice for this task as he had been a test driver in the factory for many years, and during the war was testing the tanks that the factory produced. He was one of the men destined to be part of the BMC legend. One story concerning him was told by the driver of a Porsche on an Alpine rally. He had little time to spare driving between special stages and was really trying, pushing his Porsche at high speed, when this BMC barge passed him with the driver's elbow stuck out of the window and its owner nonchalantly smoking a fag! The driver was later identified as Nobby.

The London to Sydney route took the cars across deserts and over mountains, including the notorious Khyber Pass. Crews were advised not to cross the pass alone but to join up with other rally cars to form a convoy to deter the bandits who assailed all lone travellers on the pass.

One of the desert crossings was in Iran. It was optional to take this route; the alternative was a longer mountain road. The recce had failed to determine which of these routes would be best to follow. The desert track was considerably shorter but very damaging to the cars, whereas the longer mountain road had a tarmac surface.

Tommy Eales and I were sent off to Teheran to service the cars before they tackled this section. But before this we were to be met at the airport with a hire car to recce the alternative routes. We looked at the desert crossing first. The track was worn laterally into ridges causing a washboard effect. This made for very uncomfortable riding when taken carefully, but at speeds in excess of fifty miles an hour the going felt much smoother. Unfortunately, there were pot-holes that blended in with the track surface and, by the time they were spotted, it was too late, they were taken with a spine-jarring crash.

After many miles of this treatment one of the tyres punctured. We took out the spare, which looked in good condition, but the pressure in it was very low. We were stuck in the middle of the desert with no way of inflating the tyre and were debating what to do when we saw a truck coming towards us; perhaps he could help. We flagged him down and the driver could see the problem. It was a common one in that part of the world and the truck had its own compressor to deal with it. There were two men in the truck and the driver started the compressor and inflated the tyre. To show our appreciation I got out my pack of two hundred duty-free cigarettes and gave them a packet each. The driver accepted his gratefully but his mate eyed the rest of the pack and made signs, obviously demanding them all. The driver remonstrated with him and an argument started. While this raged I told Tom to get in the car, then when he was ready I jumped in and we went speeding off down the track. We continued with as much caution as possible, but before reaching the easier going of the tarmac road the other tyre gave out; this time there was no spare. We decided that there was no alternative but to keep going. As we progressed Tom, the driver, grew in confidence. 'On this surface', he said, 'it handles just as well as it did with a good tyre.' I glanced in the door mirror - and to my horror saw black smoke belching out from the wheel. We stopped: so far there was only smoke, no flames yet, so after allowing the tyre to cool we set off again, but this time very slowly. At last, a village was in sight. We drew up at what was obviously the smithy. After a discussion carried out in sign language, and by pointing at the tyres, the smithy got the message and, from his stock of second-hand tyres, found one that would fit on the rim. Like our original, it was tubeless, but punctured. A further search unearthed an inner tube almost the right size. The next job was to get our old tyre off the rim but the bead refused to relinquish its grip - the smithy tried cutting through it with a chisel but the bead of a tyre is made of very strong steel and his chisel wasn't up to the job. By now the village population had assembled, all the men weighing in with suggestions.

There seemed nothing else for it, he would cut through it with a welding torch. He started to do this in his workshop, which soon filled with acrid, choking smoke as the tyre was burned away. He took it out into the middle of the village street to finish the job off, to the delight of the village children cheering him on. The job done, we reached the tarmac road and now had to get back to Teheran.

We went back on the mountain road and, mercifully, our tyres lasted the journey. It was a long way, and we estimated that a properly prepared rally car would make it across the desert quicker. But there was one more option. The road out of Teheran ran close to the desert, and, about a hundred miles on, before starting to climb the mountain road, there was a desert road that would meet our original route about halfway. This was the one we opted for.

The recce notes all had to be re-written for this route and all the hazards noted. In one place the track crossed a dry river bed. The bridge that had once spanned it was broken - when the BMC navigators had recced they had found that it was possible to cross this river bed by taking a track off to the left. Following their notes we took this track, only to find that a flash flood had washed part of it away and made it impassable. Turning round was impossible so we had a long way to reverse and regain the main track. We

tried the other side on foot and found this was the route taken by what little traffic there was.

The crews were doubtful about following our advice, but the navigators who had already been along there agreed that our route was probably the best option. In the event, two of the 1800s suffered suspension troubles after striking rocks or potholes, but Bill Price was 'sweeping' the route and was able to keep them going. Another casualty was the Morris 1100 crewed by three Australian girls which suffered a broken hydrolastic pipe. They were able to effect a repair with an 1800 oil cooler hose.

We returned our hired car to the owner, who was amazed that anybody should have been mad enough to cross the desert in his old Dodge Dart.

Flight to Delhi

Tom was to join Bill and Bob Whittington in the barge that they had driven the whole route from England, to continue sweeping the route down to Bombay. After delivering the hire car at the airport, I flew to Delhi for my next service rendezvous. I took off from Teheran at 11.25 that night and arrived in Delhi at 4.30 the next morning. I took a taxi to the hotel and slept until lunch. In the hotel restaurant I asked for a bottle of beer with my lunch, but the waiter was aghast; it was not permitted to drink alcohol in public. I was due to meet the local BMC man later that day so I ordered some sandwiches

London-Sydney competitors were advised to cross the Khyber Pass in convoy.
(Courtesy Paul Easter)

Works Rally Mechanic

The Golden Doored Temple of Delhi.
(Author Photo)

and bottles of beer to entertain him in my room, and when he arrived he brought his own bottle of whisky. We spent the afternoon putting India, the Commonwealth and the royal family to rights.

Later I was joined by Bill Price and his crew, who had driven the whole route from where I had left them in Teheran across the border into Afghanistan, where they snatched five hours sleep at Kabul. This had been their only opportunity to get a meal - from then on they had subsisted on the rations they carried on the barge. A hot meal was obtained by strapping cans of sausage and beans to the car's exhaust manifold which heated as they drove. From Kabul they joined the security convoy over the Khyber pass to Rawalpindi through Pakistan, as Tommy Eales puts it, 'patching up the debris on the way.'

The first car put in an appearance soon after 11.00 o'clock that night. As on many an event, the jobs were mainly routine: checking that all was secure, adjusting or changing brakes, fitting new suspension parts where needed and topping up with oil and water. After the works cars had gone through, Bill, Tommy and Bob Whittington followed them to complete the thousand miles down to Bombay without a break. They had been provided with paper underpants, supposedly so that they could frequently change them to alleviate the discomfort of their sweating bodies. Unfortunately, the sweating caused the paper underpants to disintegrate and, according to Den Green (who knows about these things), 'when you farted you blew a hole in them!'

The Australian girls' 1100 arrived soon after Bill had left, with the suspension badly down on one side. Unfortunately, with Bill had gone the special hydrolastic pump and, despite trying all manner of alternatives, I was un-

able to raise the suspension by much. Somehow, the girls made it to the boat that was to take them home. By 3.00 o'clock on Sunday morning I had seen all of the BMC cars through and thankfully took to my bed.

Later that day, after I had recovered, my BMC friend took me on a tour of Delhi, showing me the sights. The temple, with big golden doors, was very impressive, and so was the taxi ride through the streets crowded with cyclists, pedestrians and sacred cows. The cows were the only thing that the driver gave way for. The rest had to take their chances.

My rally was finished and I flew home at 9 o'clock that evening. The 1800s reached Bombay safely with three in the top ten. A major servicing operation was carried out before the cars embarked aboard the liner that was to carry them to Western Australia for the final long trek across to Sydney. Den Green accompanied Peter Browning to Australia to supervise the service crews provided by BMC Australia.

All five 1800s entered by the works finished, with Paddy in second place. They were second and third in the team prize. The third-placed team was made up by a privately-entered 1800.

The winner was a Hillman Hunter, much to the delight of the team boss: our own ex-manager, Marcus Chambers.

Tour De France 1969
Three Austin Princess barges,

plus a Triumph 2.2 PI, three rally Minis and eleven assorted personnel left Abingdon on 13 September 1969 on the journey down to Nice for the start of the Tour de France. We crossed from Southampton to Le Havre and drove to Paris, where the whole party was to take the car sleeper for the rest of the journey. We arrived at the Charolais goods depot to load the cars and found that nobody had brought the tickets; they were still in the Competitions Department office in Abingdon. A hasty phone call was made and Tony Bramley, one of Neville Challis' young storemen, was despatched with the precious tickets, delighted to have to spend the night in Paris, especially as the pretty, English-speaking travel courier agreed to show him the Paris nightlife!

The train reached Nice at 6.45 the next morning. We drove to the hotel at Cros-de-Cagnes and got established during the Sunday. Monday we checked the rally cars over and later that day the drivers took them to the first hillclimb of the event at Orme to make some notes and give the cars a final test. Tuesday was spent making final adjustments to the cars, giving them a wash and polish in preparation for scrutineering the next day.

The rally started on Thursday. Robin Vokins and I made up one of the crews in a barge. We were scheduled to meet the cars after the first hillclimb that they had recced earlier in the week.

The first service on a rally very often shows up a fault that has remained undetected until the car is driven in anger. But that was not the case on this occasion and we were able to leave for our next destination, 500 kilometres away, without hindrance.

Later on after servicing in the afternoon we were en route to service again, another 500 kilometres on, at 8.00 o'clock the following morning. Soon after getting underway we lost drive from the automatic gearbox of our Austin Princess. An inspection showed an oil leak from around the drain plug, the oil level having fallen too low for the automatic box to work. We took the plug out and changed the sealing washer, refilled the gearbox with a can of special automatic transmission oil and went on our way.

Very soon we again lost drive. The new sealing washer hadn't cured the problem, which we now identified as a faulty drain plug. There wasn't much that we could do about that and we resolved to carry on and try to find a garage with a new plug. By evening we hadn't found one and had run out of the special oil, which, according to the instruction book, was the only type suitable for automatic gearboxes. Nonetheless, we tried ordinary transmission oil - which seemed to do the job just as efficiently until we exhausted our supply of that also. Engine oil worked as well, but we were getting a little desperate.

Fortunately, we were on the route that the whole rally circus was travelling. Passing an hotel in a small town, we recognised vehicles belonging to all of the back-up reps, among them my sometime travelling companion, Ray Simpson, of Castrol. Our arms were twisted to join them in their evening meal. Then, when we explained our predicament, they were twisted even further in an attempt to persuade us to join them in staying the night in the hotel. We resisted this temptation and went on our way - with a supply of transmission oil.

By the early hours of the morning Robin and I were worn out, our stints at the wheel getting shorter and shorter as we struggled to keep our eyes open. At last we found a BMC garage. We pulled up outside, reclined the seats and fell asleep. Next morning we were awakened by the garage staff. Fortunately, they had an Austin Princess in the workshop and were quite willing to lend the drain plug from its gearbox to get us on our way.

We reached our servicing location too late to be of any assistance, but Peter was there with Bill Price and mechanic Dudley Pike. Peter was surprised to see us. 'We'd heard about your troubles', he said, 'but you needn't have worried. I made alternative arrangements.' We thought regretfully about the convivial party that we had left the previous evening, and we weren't even given the chance to feel self-righteous about it.

Towards the end of the rally we were servicing after the next-

Works Rally Mechanic

The author, with a photographic model, posing for an MG Midget poster. (Author photo)

to-last hillclimb. Paddy had suffered with a host of problems, including a broken valve spring, but was still leading his class. Julien Vernaeve and Brian Culcheth were first and second in their class, but John Handley with Paul Easter had a leaking rear main oil seal, allowing oil on to the clutch plate, and they were struggling to keep going when they hit the mountainside, causing their retirement at this late stage of the eight-day event.

Our way to the finish at Biaritz took us close to Lourdes and we were reluctant to pass this place of pilgrimage without paying a visit. I was nursing the Mini along while Robin was driving the barge with John Handley and Paul asleep in the back. They would have preferred to make straight for the finish, but Robin and I persuaded them that it would be worth spending a little time to do some sight-seeing.

The visit was extremely moving. Before we got there we were joking about what afflictions we had that could be cured. But all thoughts of joking left us when we saw the rows of crutches discarded by people cured of their disabilities in the shrine of St Bernadette. We were stunned into silence by the sight of the long procession of disabled people being wheeled in their chairs to take part in the afternoon service. Whilst there it was difficult not to believe in the healing powers of the spring waters where St Bernadette saw her vision of the Virgin Mary.

At Biaritz we had our evening meal in one of the many fine seafood restaurants. To finish off the meal we ordered brandy. I ordered a particular brand, but I saw that when the waiter passed the order to the barman he poured all the drinks from the same bottle. I sipped my drink, then called the waiter over, complaining that it was not the brand I had asked for. He blustered for a little while and then apologised, obviously convinced that I must be a connoisseur. The drinks would be on the house!

The wrong train from Salzburg

The Austin Maxi was announced in 1979. The Competitions Department was given one for evaluation and used it to recce the Austrian Alpine. Gerald Wiffen and I accompanied Peter Browning and Paddy's navigator, Tony Nash, to Austria, where the car was to be tested in the Austrian Alps. The recce didn't last long as the differential pinion wheel came adrift. We got the car to a BMC garage, where it was put up on a high lift for us to see if there was anything that we could do - which there wasn't without removing the engine. Peter decided that we should call it a day and tow the car home.

Whilst at the garage the interested proprietor came to take a look at this new car. He looked hard at me then asked me to follow him into his office where, hanging on the wall, there was an advert for the MG Midget, featuring myself and a young lady supposedly on a club rally. I could remember the poster photographs being taken some time before but had heard no more about it and didn't even know that the pictures had been used. I had never seen the poster before and have never seen it since.

The Maxi was hitched on to a tow rope and we started to make our long way home. The going was slow and Peter was anxious not to waste any more time in getting back to Abingdon. When we reached Salzburg he decided that Gerald and I should put the car on the car sleeper train and meet him the next day at the ferry terminus at the Hook o'Holland. With the co-operation of the loading crew, we got the car loaded at the siding, then went to the main station to await its arrival with the passenger carriages.

A train pulled into the platform, its destination board proclaiming it was bound for the Hook o'Holland, but it was not the train number on our tickets, and where were the cars? Confusion reigned: could there be two trains heading for the same destination so close to each other? The railway staff on the platform seemed no better informed than the rest of us. I saw the ticket collector board the train. Surely he would know. I followed him, and no sooner was I aboard when the doors were shut and we were on the move. It was not our train. I panicked 'Stop the train!' I shouted, reaching for the communication cord, but the ticket collector calmed me down. The train would stop at the next station along the line to pick up its restaurant car. I could get off then and would be taken back to the station. A railway worker escorted me from the train to a siding where a locomotive was just leaving and I was invited on to the footplate. The driver was in no hurry to move off and I anxiously impressed on him that I might miss my train to the Hook o'Holland. He grinned, 'I drive the train to the Hook o'Holland with this engine', he said, 'you won't miss it!' Gerald and the rest of the bemused passengers had last seen me disappearing into the distance, apparently on my way to Holland. They were amazed to see me reappear standing on the footplate of the engine.

The car used on the rally was a new arrival and heralded the start of the Triumph takeover of the department: it was a 2.5 PI MkI. The Triumph Competition Department recommended modifications that should be carried out on the car, which were confined to the brakes and suspension.

Paddy wasn't too happy with the gear ratios and, after pre-rally testing, asked us to change the overdrive wiring, to allow the overdrive to operate on all four forward gears. We achieved this by by-passing the inhibiting switch on the gearbox, thus giving overdrive on all gears, operated by the manual switch on the gear lever. Then, to obviate the possibility of the overdrive operating on reverse, which would have broken the gearbox, we used a cut-out relay wired to the reverse light switch.

This new system proved too much for the clutch, which failed towards the end of the rally. There was no time to change the clutch and Paddy retired from the event. The car was then sent home on the train - without any dramas.

The Triumph Competition Department had been in existence longer than its BMC counterpart, and its team can't have been very pleased that all of their work was to be taken over by Abingdon.

The World Cup
In 1970 the World Cup football finals were held in Mexico, and a rally was organised to visit every

DAILY MIRROR WORLD CUP RALLY 1970

Royal Navy   Mr. R. Redgrave   Metropolitan Police   Peter Jopp   WOMAN Magazine.

country that was represented in the finals.

All the major manufacturers entered teams of cars in the event. The BL main assault was by four Triumph 2.5PI Mk2s, crewed by Paddy Hopkirk/Tony Nash/Neville Johnson; Andrew Cowan/Brian Coyle/U.Ossio; Australians Evan Green/Jack Murray with Hamish Cardno; and Brian Culcheth/Johnstone Syre. One other Triumph was privately entered and driven by Roy Fidler.

A crew of pilots from the RAF aerobatics team The Red Arrows (Terry Kingsley/ Peter Evans/ Mike Scarlett), drove an Austin Maxi, as did Rosemary Smith with Alice Watson and Ginette Derolland.

Five 1800s were prepared by Special Tuning for private entrants and a team of Special Tuning mechanics was sent to look after them.

The cars prepared by Special Tuning for the World Cup Rally of 1970, and the mechanics who built them. The photograph has been signed by the crews. (Courtesy Basil Wales)

Europe

The rally started from Wembley Stadium on 19 April. 1970 but, by then, we were well on our way to our first service. A cavalcade of five cars left Abingdon to cross from Southampton to Le Havre. I was sharing a Rolls-Royceengined Princess barge with Gerald Wiffen and the Department's panel beater Stan Bradford. We travelled in convoy to Yugoslavia, passing through the Mont Blanc tunnel on the way and spending the first night in Aosta. In Yugoslavia we dispersed to our various destinations.

We were headed for Titograd - not my favourite place - where we were to spend two nights at an hotel called the Hotel Crna Goro, which was every bit as horrible as it sounds. My bed had a hole in the middle that I slowly slid through during the night.

The next day our rally started. The Triumph 2.5 PI driven by the Australian, Evan Green, had gone off the road on the previous stage, damaging the car's front suspension, and we fitted a new front strut.

John Handley's Mini arrived after coughing and spluttering its way round the stage. The trouble was obviously caused by fuel starvation so we changed the pump and stripped the carburettors before discovering that the actual cause was the supplementary fuel tank that had been fitted in addition to the two that were normal on a rally Mini. This had called for a complicated job of plumbing that, in effect, didn't allow petrol to flow freely from this third tank. We advised John to run with more petrol in the normal tanks until time could be found to rearrange the pipework.

Roy Fidler, in his privately prepared Triumph, had been breaking rear shock absorbers. An inspection showed that a protection plate on the bottom of the

Mike Scarlett with Red Arrows pilots Terry Kingsley and Peter Evans, who drove an Austin Maxi on the World Cup Rally.
(Courtesy Basil Wales)

Works Rally Mechanic

World Cup Rally of 1970. John Handley and Paul Easter in the Mini Clubman which failed to finish the European section. (Courtesy Mick Hogan)

The Mini was still misfiring and had been diagnosed with a burnt piston brought on by fuel starvation. It was destined not to reach the end of the European section of the rally.

The Evan Green car was in trouble again; his engine was misfiring and an earlier crew had

'shocker' was incorrectly fitted and was preventing the 'shocker' from realigning to follow the suspension movement.

Meanwhile, all the rest of the team cars had been through and it was time that we were moving. The next leg of our journey was to be made by plane, so we returned to Titograd and left the barge with Derek Plummer who, with Castro I's Ron Stacey, was to drive it home.

We boarded the plane, but take-off was delayed because the last person aboard found that there wasn't a seat for him; the plane had been over-booked. He was a large, aggressive man and defied all efforts by the crew and police to eject him. Eventually, he won and a woman was forced off the plane in his place. She wasn't pleased.

Our destination was Nice. A hire car awaited us at the airport, in which Stan Bradford,

Extra petrol tank improperly plumbed, causing the Handley/Easter retirement. (Courtesy Den Green)

new companion Gordon Bisp and I drove to Ventimiglia on the French-Italian border, where we met the two Tommies, Wellman and Eales, who had been servicing on their own schedule in another of the Princess barges. Together, we would be servicing at Camporosso the next day.

changed injectors in an attempt at a cure, but without success. When he reached us we were able to give the engine a thorough checkover. I adjusted the tappets and found that the number six exhaust valve had a lot of unusual movement. By putting a screwdriver through the sparkplug hole we were able to hold the valve in place while we removed the valve spring. With no valve spring blocking the view, we could see that the valve guide had disintegrated. The bits must have been blown out of the exhaust manifold.

The only cure was going to be a new cylinder head, which we didn't carry, but the next crew did so it was a matter of making the car roadworthy again by replacing the valve spring. The 'floating' valve was likely to break, causing irreparable damage to the engine, if it continued to be operated, so we removed the push-rod and blocked off the fuel feed to that cylinder. In fact, it proved impossible to get the cylinder head changed and Evan Green continued with his five cylinders to the end of the European section at Lisbon.

Roy Fidler was another customer with misfiring problems caused by fuel starvation. He told me that the previous mechanics had changed the feed pump, the high pressure pump and the injectors, all to no effect, which left only the most awkward bit, the metering unit. I fitted a new unit but it made no improvement. Roy was already running late and was eliminated for being outside of his allowed lateness at the next control. He later told me that a subsequent inspection after the rally showed that it was the new high-pressure pump that was faulty.

Tommy Wellman, Tommy Eales and myself returned to Nice airport in the hire car, leaving Stan Bradford and Gordon Bisp to drive the barge back to Abingdon that Tom had arrived in. Our control at Camporosso had closed at 11.45pm and we were on the plane to Lisbon at 10.35 the next morning.

We landed at Lisbon at 1.55pm and went straight to the garage of Goncalves where Tommy Wellman and I picked up a car, with driver, to take us to service at the end of the last special stage of this European leg. Our driver lived in a village 150 kilometres from Lisbon en route to our destination. He had been chosen for the job because it would give him a rare opportunity to visit his family. We were given a warm welcome in his village and were plied with coarse red wine and knobs of goats' cheese. Eventually, we left him with his relatives and drove the remaining 50 kilometres to our service point at Pampilhosa.

We were here to see the cars off the end of the stage and to make sure they were fit to drive down to Lisbon. The first car was due at 3.30am and they continued coming during the course of the night and morning until, at last, there only remained the Triumph of Evan Green.

Evan's troubles had pursued him all the way, and his lateness had forced him to drive with less caution than he would have liked; consequently, he crashed the car and limped into Bill Price's service extremely late. Bill was at the start of the stage that we were on the end of. He changed the damaged parts of the car, and when Evan reached us he stopped for a very short time, anxious to reach Lisbon where the cars were to have a thorough overhaul.

We left the control at 2.30 that afternoon and, after picking up our driver on the way, slept until we reached Lisbon. When we arrived it was to find all the lads tackling the various maladies afflicting the Evan Green car. At last it received a new cylinder head, all new suspension parts and brakes, its broken lights were changed and the bodywork straightened out.

All the cars received similar treatment, and we all suffered from lack of sleep, so it was with great relief that the cars were locked away in *parc ferme*, where we were not going to be able to get at them again until they reached South America.

South America

The whole back-up team took off from Lisbon in a British Caledonian Britannia. Three-quarters of the seats had been removed to make space for the spares that were going to be spread around South America. We stopped off to refuel at Sierre Leonne, where the health authorities boarded the aircraft and sprayed it and us with a foul-smelling disinfectant. After this we were allowed off while the plane was refuelled, but, as a tourist attraction, the airport building left a lot to be desired and we felt that we needed disinfecting again before we

Andrew Cowan's World Cup car being serviced at Lisbon. (Courtesy Mick Hogan)

Alice Watson, Ginette de Roiland and Rosemary Smith chat while their car is serviced at Lisbon. (Courtesy Mick Hogan)

re-joined the aircraft.

We who were left in Rio put the time to good use, taking in all the delights of the city. During the day much time was spent on the beach at Iponima, where Terry Kingsley, the Red Arrows pilot, suffered a bad experience. While he and his companions were swimming, someone stole the wallet which he had ill-advisedly left unattended with his clothes. For the next couple of days Terry had us mobilised. One pair was detailed off to swim, leaving their belongings on the beach, while the rest of us innocently lounged at various points surrounding the beach, waiting for the thief to strike again. Nothing happened, of course. I think that the little knots of sunbathers with skin shades varying from pasty white through pink to lobster red were a little bit too conspicuous.

One highlight was a trip to the Maracana Stadium to see Pele playing for Brazil against Austria in a World Cup warm-up friendly. Our whole crew was taken in a hired bus. We were all wearing our blue World Cup Rally jackets and were mistaken for the Austrian team as our coach threaded its way through the throng of jeering and cheering Brazilian supporters. The stadium is huge and we were amazed at the noise level - everybody had a transistor radio and followed a commentary of the match they were watching. The din was unbelievable.

After we landed at Rio de Janeiro we had a wait of a week for the rally cars to be shipped over, but Bill Price and some of the mechanics flew on the next day to distribute spares and personnel to strategic points around the route.

Armed escort

The rally was due to re-start on 8 May. I was teamed with a great friend, Johnny Evans, who had started working in the factory directly on his release from National Service. John was a Devonshire lad who had been serving near Abingdon at the time of his demobilisation and was courting a local girl. Rather than return to his home in Devon, he got a job in the MG Service Repair Department, working on the next bench to me until I moved to Competitions. John followed me after Service Repair was closed and he had spent some time in Special Tuning. Until he married, my wife and I had John as our lodger.

We were to leave Rio with plenty of time to spare to get to our first service point. Our first instruction on the schedule was: Monday 4 May, go with local driver in truck with spares and tyres to Canela (end of Prime 8) approx 700 miles. 'Prime' was the term used instead of 'special stage' and some of them were in excess of 500 miles, all to be driven at racing speeds.

We picked up our truck from the local BMC agents and were introduced to our escort, whose name was Eugene. Instead of driving straight out of the city Eugene drove the truck to a nondescript building where he motioned us to stay in the truck while he went inside, re-emerging with a package which he slipped into the glove compartment.

Our way to Canela was by way of a main road just inland from the South Atlantic coast. On the outskirts of Rio on the main road, there was a fortified roadblock at a diversion we were only allowed to take after our credentials had been inspected.

The diversion led us further inland and was over a mountainous road skirting Sao Paulo. At frequent intervals we were halted at more roadblocks, sometimes with a queue of traffic waiting a turn to be allowed through. Eugene would drive to the front of the queue and was given priority treatment. We asked him how he managed it and he grinned. 'There is much bandit activity in the area', he said. 'I am a special police officer and I have been given the job of getting you safely to your destination.' He showed us his warrant card and, opening the glove compartment, took out a revolver. 'I hope I don't have to use this.' (We fervently hoped so as well!).

The frequent stops and poor roads meant that, with our laden truck, we had not got as far as we would have hoped when we stopped for the night. The next day we fared a little better, being allowed back on to the main route, but still with the frequent roadblocks. That night we stopped in a small town where there was a decent hotel. After getting cleaned up, we followed Eugene to where he said we could get something to eat and drink. He led us to a dingy wooden building. Inside was a bar. The males in the place were the very epitome of South American bandits who could have stepped straight off a film set. The females completed the illusion. We sat down at a table and were promptly joined by three of these lovelies. We had a surprisingly entertaining evening with Eugene acting as interpreter, although they did have a limited grasp of English with a heavy American accent. However, the atmosphere changed when, soon after midnight, we made it obvious that we were leaving to return to our hotel without them.

The barman demanded a ridiculous amount of money for commission on the drinks that the girls had consumed. Eugene refused to pay; John and I furiously searched our pockets to see if we had enough to get us out of the place alive. One of the regular customers had actually taken out a knife and ostentatiously laid it on the table in front of him. But Eugene would have none of it and he eventually showed his police warrant card. This was enough to get us out, only to be escorted to the local police station by all the hostile bar patrons.

The police station was shut off by a big iron gate behind which was a uniformed figure who pointed a gun through the gate and told us to clear off - even we got the gist of that. We obviously had money, and our enemies were intent on getting their hands on some of it. The man with the knife was still with us. Eugene led us to the hotel; on arrival he told us to go straight up to our room. Once there we pulled the dressing table against the door and wedged it with the bed.

The next morning we were relieved to see Eugene alive and well. He hustled us out of the hotel and on our way without stopping for breakfast. We asked how he had got on in the night. 'I had visitors', he said, 'but I showed them my gun and they went away again.'

Tom and Dudley in Montevideo

Tommy Wellman and Dudley Pike were one of the crews that stayed aboard the Britannia to be dropped off at Montevideo with a supply of tyres and other spares. It was one o'clock in the morning when they landed, and with a brief stop to unload Tom, Dudley and the spares, the plane took off again to continue on its way around the rally route. Unfortunately, the customs officers on duty knew nothing about any arrangements for bringing these parts into the country and the two mechanics were whisked off to the nearby police station where they were interrogated until nine o'clock the following morning, when the British Consul was contacted and was able to gain their release and sort out the customs details.

From Montevideo, Tom and Dudley had to travel, in a truck with a local driver, to the end of the 'Prime' at Saito in Uruguay. Here, they were met by an agitated Peter Evans of the Red Arrows crew. They had been in a collision with an ancient Ford Popular, written it off and damaged the Austin 1800's suspension. Tom went off to the scene of the accident and found the lady driver of the Ford mourning the death of her dog, killed in the smash. After some negotiation she became reconciled to her loss and happily accepted the Uruguayan equivalent of £200 that Tom was able to pay her.

In Montevideo the service crew got a night's rest in an hotel that overlooked the River Plate. At low tide the German battleship, scuttled in an engagement with the Royal Navy in World War 2, was plainly visible. They were woken in the morning by the rumbling of tanks, as the army carried out a coup.

The Red Arrows 1800 was soon made roadworthy again and the hydrolastic suspension pumped up with the car's own emergency pump. When they reached the service at Montevideo a new driveshaft was fitted. Damage to the battery showed up later and this was replaced by Bill Price's crew at the next stop.

A Chilean epic

After our adventures on the journey through Brazil we finally

Emergency hydrolastic pump carried by the works cars.
(Courtesy Peter Browning)

reached Canela where we were due to be met by Bill Price, with Gerald Wiffen, Den Green, and Peter Browning, who were to have made their way to us by Cessna. But a major job on Andrew Cowan's car - changing his overdrive - had delayed them. Andrew had also packed his passport in with his personal luggage, and Bill had to retrieve and get it to Andrew before the cars crossed into Argentina. So John and I were left to our own devices.

Doug Watts and Robin Vokins had been at the start of the Prime with an emergency spares kit, so their servicing was confined to routine checks. We had a full set of spares with wheels and tyres and were kept busy changing these. This was to be a job at most of the main controls - there were 370 wheels distributed around South America and they would all be needed.

After seeing all the cars through we drove the fifty miles to Porto Allegre. Here we parted company with Eugene and caught a plane to Bueno Aires, staying the night there in the City hotel.

The next leg of our journey took us right across Argentina to Santiago on the Pacific coast in Chile. We took off in the early morning and were in Santiago before midday. Here, we were met by representatives of Leyland Chile, who provided us with a truck loaded with spares. Peter Browning had got there before us. I had been studying our schedule and told him that I thought it was pushing it a bit to get to our service point and back to Santiago in the time allowed. Peter, though, was quite confident. 'Don't worry, you'll make it', he said. It was originally intended that there would be only one local driver, but we were going to have to travel 600 miles, halfway to Cape Horn, to service

Refreshment halt *en route* to the service point in a local pick-up truck.

at the town of Puyhue on the Argentine-Chile border. The first rally car would reach there at 3.00pm the next day, so the Leyland agent decided that we should have two local drivers. John and I were not too happy about this as the transport was an open pick-up truck and four of us sharing the cab was going to make for a very uncomfortable ride.

This was another occasion where the drivers passed through their native village and they planned to make the most of it. We were obviously expected: the first Englishmen that many of them had ever seen. A buffet meal had been laid out in the

village hall and the local wine was dispensed with great generosity. A fiesta atmosphere prevailed, and we left the party with reluctance: it would continue, even though the excuse for it had left.

Our drivers had enjoyed a few bottles of wine, and it was soon obvious that they were not going to be fit to get us safely to our destination. Consequently, John and I shared the driving for the rest of the way, with one of the drivers asleep among the spares in the back of the truck and the other one curled up in the corner of the passenger seat.

We reached Puyhue after travelling all Monday night and Tuesday morning. The first car

was due there at 3.00pm and rally cars would be crossing the border until the early hours of the morning.

Great preparations were in hand. This was a major customs post and the customs officers were quartered in houses near the border. We were invited in to the Chief Customs Officer's house, where all the rally officials were being entertained. By the time the cars arrived, the evening had really got underway with a band and a line of South American beauties dancing and showering the cars with flowers as they crossed into Chile. What seemed like herds of cows were sacrificed and roasted over the huge barbecue pits and the business of carrying out necessary repairs took second place.

To get back to Santiago on schedule meant leaving straight after our last car had been through. Our gruelling journey down, and our poor attempts to fight off all the offers of wine and beef sandwiches, made this a definite no-no.

We had been introduced to the leading land owner of the district and had been to his ranch where he promised to present us with an animal skin before we left. We imagined, perhaps, a mountain lion skin or some such, but it turned out to be a little calf hide that we soon discarded because of the smell. Nevertheless, Señor Dobra Dobrovski, for that was his name, was a very rich man. We discussed our trip and he agreed that it was too long a

journey for us to undertake without a night's rest.

We decided that the only way we would make it would be by aeroplane. There was a small airfield where we could catch the daily flight to Santiago. It was far enough away for Snr Dobrovski to fly us there from his ranch the next morning; meanwhile, he arranged a room for us in an hotel for the remainder of the night. We sent the truck and the drivers away to make their own way home.

The next morning was too foggy to take off so Snr Dobrovsky sent a truck with a driver to take us to the airfield. It seemed that all Chilean truck drivers are daft, and this one was dafter than most. He simply assumed he had the right of way everywhere: major roads, traffic lights ... he treated them all with equal contempt. To avoid one collision he had to make an emergency stop, and the brakes grabbed on the nearside, sending us into the shallow ditch alongside the road. He was able to drive out again and continued on his way with the same gay abandon.

The flight was delayed while some of the seats were taken out to make way for a coffin whose occupant was to be buried in his home town of Santiago.

We took a taxi from the airport to our hotel where we were met on the steps by Peter. 'Ah, here you are. I told you you'd make it easy enough, didn't I?' he said ...

Across the Andes by Cessna

We left Santiago the next morning in a front and rear engined Cessna. The passengers were intended to be John and me, Peter and Den Green. At the last moment, however, Graham Robson hitched a lift. He was to be in charge of the control at Ovalle where we were dropping Peter off. Graham had a huge sack full of the equipment and paperwork he would need at the control. There wasn't room for all of us and our luggage, so John and I had to leave our suitcases to be taken on by road. It was in this confused situation that I lost my camera with all the unrepeatable photographs I had taken in South America, particularly of our policeman, Eugene, with his truck, and the celebrations at the Puyhue border.

Our pilot on the Cessna was called Carlos Diaz. He told us that Diaz was Spanish for 'days' and his friends made a great joke about his seven children, calling them his 'week'. We took off to cross back into Argentina, landing at the border to allow Peter and Graham off before we tackled the mountains of the Andes. We were faced with a strong head wind. I was looking down on the same bit of rock high in the mountains for what seemed like an eternity; we were just not making headway. Carlos had to veer off and make a circuitous route to San Juan, where Den, John and myself took a hire car a further 150 miles on to service

Blinding dust clouds followed the cars in South America.
(Courtesy *The Daily Mirror*)

Andrew Cowan's car after his crash on the World Cup Rally. (Author photo)

at Rodeo with an emergency kit of spares.

Evan Green had been forced to retire by now - the piston in the misfiring cylinder had been damaged by the ordeal and had disintegrated. All the rest of the works cars were still going strong, although the Triumphs were proving no match for the Ford Escorts, who were occupying the first four places in the rally, with Paddy and Brian Culcheth fifth and sixth. The Maxis were having a hard time of it. Terry Kingsley had been in a collision and had also needed a driveshaft changed, but Rosemary Smith was leading the ladies.

We drove back to the Cessna and flew to the middle of the fourteenth 'Prime'. It was here, in the middle of nowhere, that the Ford Escort driven by Rauno Aaltonen broke an engine mounting. The Ford mechanics had taken it off and welded it but the oxy-acetylene welding had not been strong enough, and here it was, broken again. There seemed to be nothing more they could do. It needed electric welding, but the Ford mechanics gave up. 'We won't find an electric welding kit in this god-forsaken hole', they said, and were astounded when, out of the crowd of onlookers, a man stepped and said in perfect English 'Come with me. I have an arc welder at my farm.'

We left soon after midnight and were at Salta, the end of the Prime, before the cars got there at 12.40. Soon after we got to

Salta we had news of a serious accident to one of our cars. Andrew Cowan caught the privately entered Maxi driven by Jean Denton and overtook her on a bend. The bend took him in the direction of the rising sun which, shining on the impenetrable dust cloud raised by the Maxi, completely blinded him. The car left the road, which was twenty feet above the surrounding countryside, and landed on the roof. The crew of the following car stopped to tend to Andrew, who had sustained a cracked vertebra, and sent word to Salta for an ambulance and a doctor. Soon Andrew's crew was in hospital in Salta. One of the crew of the following car was a journalist. He gave us a press release to wire to his paper with the news that Andrew had been killed in the accident. John and I went to the hospital and found that, although Andrew had serious injuries, they were not expected to prove fat al. Andrew's two crew members had sustained head injuries but would soon be on their feet. We were glad we had checked for ourselves before spreading the sensational rumour.

Directly after seeing the works cars through we left Salta to fly the 600 miles to La Paz, capital of Bolivia, high in the Andes. On the way the navigation system gave a little trouble. Flying over featureless mountain tops, Carlos was taking his directions from signal beacons which told him where he was, and his chart showed him the compass

heading to the next beacon. For a long interval there was no bleep-bleep from the radio. Carlos tapped it several times without getting any response. He shrugged philosophically. The beacons were unmanned and sometimes those in remote mountain locations failed through lack of maintenance, he told us. 'No problem, I know the way', he said.

Our pilot would start his flight using fuel from the reserve tank until this was down to half level. He then turned to his main tank and exhausted that before turning back to reserve. I was sitting up next to him and I realised that he had let his reserve run lower than usual before turning to main tank. When the time came to run on reserve I could tell that he was shocked to see how little he had left himself to complete the flight.

We were having to fly higher now and needed to use oxygen. There were no oxygen masks, but a bottle with a plastic tube that we each took a suck at as we felt the need. Eventually, the bottle ran low and Carlos had first call on it; the rest of us had to suffer. We should have approached La Paz from the south but Carlos had used Lake Titicaca, which was north of La Paz, to get his bearings. Carlos explained that it was forbidden to fly over the lake but as we were low on fuel and oxygen he would risk the wrath of the lake's spirits and head straight in. We circled the airfield, asking for landing permission, but got no response at all from the control tower. In despair, Carlos said 'We shall have to land. Keep your eyes open for any other planes landing at the same time!' Fortunately, the airspace and landing strip remained clear. After landing, Carlos went to the airport authorities to remonstrate with them, but was fobbed off with excuses. Carlos was of the opinion that the lone traffic controller had dozed off!

Our service was at the airport. Working here at an altitude of well over 12,000 feet was exhausting. This was a major service area and a night stop, so there was plenty that we could do to repair accident damage and replace tired suspension parts. There was a medical centre and we frequently had to break off work to go there and take a whiff of oxygen. Paddy arrived late after having a differential changed some miles back down the road. His lateness gave us less time to work on his car and, consequently, less opportunity to visit the medical centre. As a result, we were suffering badly from lack of oxygen before we finished. That evening, before gratefully getting to bed, we arranged to borrow two oxygen bottles from the Cable and Wireless Company, to get us over the mountains to Cuzco the next day.

This was destined to be our last flight in the Cessna. After leaving La Paz with full fuel tanks and a ready supply of oxygen, we thought that our troubles were over. But soon the front engine started to misfire. Carlos told us not to worry; he assured us that the rear engine could keep us at an altitude of 10,000 feet. I didn't take much comfort from this, however, because the average height of the peaks we had to cross was more like 16,000 feet!

We landed at a grass airstrip, lined with huge cactus plants which towered above the plane. Carlos and a mechanic from a workshop at the airstrip diagnosed the problem as being with the fuel supply and soon had it fixed. We took off again, surrounded by mountains. The only way the plane could get enough height to get out of the valley was to spiral upwards. Thankfully, the engine was running cleanly now.

Still in my seat next to the pilot I was constantly checking all the dials in front of me. The oil pressure on the previously troublesome engine started t9 fall. I nudged Carlos and drew his attention to it. The pressure continued to drop until he had to cut the engine and we proceeded to Cuzco on one engine, dodging the higher peaks. We made it, and when we landed found that the whole of the underside of the aircraft was smothered in oil. It seemed that vibration caused by the earlier misfiring had fractured an oil cooler pipe.

We serviced here with Tommy Wellman and Dudley Pike, who had arrived by truck with spares. After servicing, Gerald Wiffen, Johnny Evans, Den Green and I were supposed to take the Cessna on to Lima, but the Cessna wasn't going anywhere without its oil cooler pipe, so we carried on to Lima by scheduled airline.

Tom and Dudley were able to stay in Cusco for a couple of days and took advantage of this by taking a trip to the 'Lost City of the Incas', which had been discovered in 1939 by an American called Sanderson. To get there they took a ride on an ancient steam engine that pulled two coaches, zig-zagging first forward then reverse, up a mountain. Their way led through a narrow cutting dug out of the soft sandstone. A gang of men travelled in the front coach: their job was to shovel away the loose sand that fell from the walls of the cutting and blocked the line. After a six-hour journey they arrived and were amazed at the precision of the huge blocks of stone that fitted together to form the arches and walls of the ancient buildings. The only permanent inhabitants were wild llamas. The city was at such an altitude that Tom and Dudley had to use oxygen to give them the energy to walk around the ruins.

The new oil pipe arrived the next day and when it was fitted they flew on in the Cessna to their next control at Cali in Colombia.

Special Tuning deserted

Lima was another major control and we were there in force. Besides Bill Price and Peter Browning, there were Gerald Wiffen, Den Green, Johnny Evans and myself, plus two Special Tuning mechanics, John West and Martin Reade. The field was well spread out now. The first car came in at 9.00pm, and at 6.00 the next morning we were still waiting for two of the Special Tuning prepared cars to turn up. Peter decided that there was no point in us all staying any longer and the Competitions Department mechanics should get some sleep, leaving Martin and John West to look after their cars when they arrived. Martin and John were not too happy about this, having worked all night with us on whichever car turned up, and I must confess that I felt reluctant to leave them to carry on alone, but I was too tired to volunteer to stay and help them.

Stranded in lion country

The Cessna left Lima loaded with personnel. Tommy Eales and Roy Brown were en route to service close to the Peru-Brazil border.

On the way they touched down to let another service crew out at their appointed place. The pilot took advantage of this stop to refuel. The fuel came out of rusty old churns with a piece of muslin stretched over the filler to catch the worst of the rust particles. They also took on board a local who was to act as guide.

The plane approached the destination but there didn't seem to be a landing strip. They circled the area, then decided to land on a likely-looking piece of terrain. After dropping Tom and Roy, the

Aircraft refuelling, South American style. (Author photo)

Tommy Eales lost in lion country ... (Courtesy Tommy Eales)

plane took off on another schedule which would bring it back in four days' time to pick them up. Dusk was approaching and Tom and Roy appeared to have been left in a barren area of countryside, although a village had been spotted before they landed. The guide set off to find the village, leaving Tom and Roy with his luggage. The evening passed with no sign of the guide returning.

Two trail bikers came roaring along and, in sign language and pidgin English, advised them to make for the main road, which was in a different direction to the one taken by their guide. They took the bikers' advice and reached the road just as two Land-Rovers drove up. The drivers were English-speaking locals who had seen the plane come down and assumed that it had crashed. When Tom and Roy recounted their tale their rescuers

told them that it was as well that they had found them before dark because that was when the fierce mountain lions started hunting!

Back in the village it was explained to them that they had landed the wrong side of the border and would have to go through immigration procedures before

they could recross the border to their destination about two miles away. They reached their hotel eventually, where they were amazed to see their bedraggled guide arrive in reception. He was without a passport and, rather than risk an encounter with the border officials, had found a way

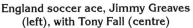

England soccer ace, Jimmy Greaves (left), with Tony Fall (centre)

to crawl through the barbed wire fence.

Four days later and the Cessna was due to return. This time they lit bonfires to ensure that the pilot would find the right landing place when he got there.

Trouble in a banana plantation

The rally still had a long way to go, but Brian Culcheth was lying second to Hannu Mikkola and

Brian Culcheth and Johnstone Syer on their way to second place in the Triumph 2.5 PI on the World Cup Rally. (Courtesy *The Daily Mirror*)

we felt satisfied with the way it was going. The cars were to be shipped aboard a liner from Colombia to Panama, affording the crews some time for relaxation. During the three-day voyage, Brian Culcheth and Johnstone Syer celebrated the birthday of Jimmy Greaves, the English soccer star, who was competing in the rally with Tony Fall in a Ford.

Meanwhile, we took yet another flight - this time our destination was San Jose in Costa Rica. We were joined here by Gordon Pettinger, the Dunlop tyre fitter. The local Leyland representative met us and provided us with two Land-Rovers: one for John and me (with one of the kits of spares that Bill had distributed) and the other loaded with tyres for Gordon. We had to travel 200 miles to our service point.

That night we stayed in a small motel. There was no eating facility in the motel but it was right by a banana plantation and the motel proprietor, who was an expatriate American, directed us to the plantation workers' canteen. This was the social centre of the plantation. It was Saturday night and there was a dance in the canteen. After our meal we were invited to join in the evening's entertainment. No alcohol was allowed in the dance hall, so we made frequent trips out to the bar until the barman filled up some Coca-Cola bottles with brandy and told us to take them away.

Soon the long drive and the brandy caught up with me and

I suggested we called it a night. Gordon persuaded John to join him in one more drink, and I made my way back to the motel in our Land-Rover. The next morning Gordon was anxious to get started. As we drove John told me of the happenings of the previous night after I had left. They had a few more drinks and Gordon started chatting up one of the pretty señoritas, much to her boyfriend's annoyance. A major row developed and the crowd turned nasty, chasing after them as they took off in the Land-Rover. Gordon pulled into a driveway and turned off his lights. Their pursuer sped on into the night, and Gordon drove back to the motel the way he had come.

Mexico at last

After our stop at Canoas we returned to San Jose where we joined up with Doug Watts, Gerald Wiffen, Robin Vokins and Martin Reade. Our rally was done. The flight to the finish in Mexico City turned into a party as we started to wind down after three weeks of chasing the rally practically the whole length of the continent.

We took taxis to our hotel, the Camino Royal, and from there were ferried out to see the cars arrive at the finish in the Aztec Stadium. Out of the ninety-six cars that had left Wembley six weeks before, just twenty-three made it to the finish.

Brian Culcheth took second place, Paddy fourth, Rosemary Smith tenth (and the Ladies' prize), Red Arrows' Terry Kingsley was twenty-second. The other three works cars had perished en route: Evan Green and John Handley with engine troubles and Andrew Cowan in his near-tragic accident.

Due to a faulty windscreen seal our plane was being repaired in Montreal, which threw the return journey itinerary into disarray. Instead of taking us straight home, our new journey would entail flying across the border into the USA where we changed planes for a flight to Montreal.

The airport in America was equipped with a band for entertaining. It wasn't playing while we were there - that is, until Peter Browning seated himself at the organ and Paddy at the drums. Both are accomplished musicians and their performance was applauded by all the waiting passengers.

Our arrival at Gatwick caused a stir when Doug Watts seated himself on the luggage carousel with two more of the lads all wearing big sombreros, while we gave an impression of the theme music of *Sunday Night at the London Palladium*.

The chop

Later on in the year, after Peter Browning had tried unsuccessfully to wring from our Leyland bosses details of their long-term plans for the department, we were given the news that it was to close.

Initially, MG management was adamant that the whole workforce would be made redundant. The unions were unsympathetic: their members saw us as having had a privileged position in the factory and we had taken no part in car production.

The final outcome was that some found work of some kind around the factory, whilst others took redundancy. Robin Vokins and Mick Hogan moved to Essex to join the Ford Competition Department under their old boss, Stuart Turner. Robin was still there in 1997 as Workshop Foreman. Bill Price left to take a position as Service Manager to a BL garage in High Wycombe. The supervisory staff were made production line foremen. Three of us - Gerald Wiffen, Eddie Burnell and myself - were transferred to Special Tuning.

Basil Wales

Working with the MG staff at Abingdon was an experience that I am sure none of us would have wanted to miss. Brief association during my apprenticeship at Morris Engines had whetted my appetite and my job as Sports Car Specialist brought me into contact with the enthusiastic management. The offer of the job of Special Tuning Manager was an opportunity I hadn't expected, but took on with relish. Stuart Turner proved an excellent boss who never looked over my shoulder but was always forthcoming with advice when needed. Normal Higgins was delighted when I turned loss into profit and the staff helped me generate a team spirit that even encompassed family members when out rallycrossing at weekends.

The inter-dependence of each mechanic, car crew and management is vividly demonstrated in Brian's anecdotes. Major decisions were taken during events by individuals without the modern benefits of blanket radio communications and helicopters!

Brian's very readable story gives yet another insight into the Abingdon experience which we are all proud to have shared. What other workplaces have been the subject of so many interesting books? I guess no others deserved it!

Basil Wales

Basil Wales

Special Tuning

In 1964 the MG Service Department was closed. Bill Lane - the workshop foreman -, and half of the fitters were dispersed around the factory, but Bill Burrows and the remaining fitters formed a hard core of skilled men that John Thornley was loath to see going to waste.

This was the year that Paddy Hopkirk won the Monte Carlo Rally and the Competition Department was inundated with calls from Mini owners asking how they could make their Minis go like Paddy's. Bill Price, in the Competitions office, was unable to answer all of the requests and, after further discussion, John Thornley announced the formation of a new department, Special Tuning. The name was chosen because Syd Enever had already been writing notes on tuning the MG under the heading of 'Special Tuning'.

The Cooper racing team was using BMC engines in its Formula Junior cars and servic-

the new Special Tuning Department. The bulk of the information sent out by the department was on which parts were needed and how to modify and manufacture parts to enhance a car's performance. Basil saw that in do-it-yourself shops, kits were on sale to do complete jobs and he applied this principle to make up tuning kits. This simplified the tuning of cars in as much as a kit would be provided with the special parts, complete with the nuts and bolts to do the job, all under one part number. This, then, was the basis from which Special Tuning grew.

Rally car preparation

There was also the facility for a

Basil Wales, leaning on the table, taking over Special Tuning from Glynn Evans on the right.
(Courtesy Basil Wales)

Special Tuning parts display.
(Courtesy Basil Wales)

ing these was the first job undertaken by the department. Glynn Evans was in charge, but his father - who owned a garage business in Wales - was anxious to hand over the reins and, in due course, Glynn left the factory.

Basil Wales had Sportscar Specialist Service Engineer, travelling the world sorting out BMC sportscar problems. In this position Basil was well-known to the Abingdon Service, Quality and Design staff and to John Thornley, who invited him to lunch and offered him the job of managing

The Special Tuning workshop.
(Courtesy Basil Wales)

private, would-be competitor to bring his car to the factory and have it prepared by experts. The ex-service fitters took to this so well that, in 1966, Basil was asked to prepare a works Mini Cooper S for the Scottish Rally for Tony Fall to drive with Mike Wood. Two Special Tuning mechanics accompanied Basil to Scotland to service the event; the only non-routine job was changing the rear wheel bearings and the car went on to win the event.

Sitting in their support car in the middle of the night waiting for the cars to arrive at a service point, Basil and his mechanics were interested in the progress being made by a lady with a tray, who was having a word with each of the other service crews. Eventually she reached Basil and asked if this was the BMC Ser-

vice crew. When Basil assured her that it was, she told him that she and her family had been so impressed the previous year by the sound of Timo coming over the mountain in a Healey, they had become firm BMC fans and had brought them some tea and cakes!

The department also pre-

Basil Wales and his mechanics servicing
on the Scottish Rally.
(Courtesy Bill Burrows)

pared a 1275 Cooper S engine for a French driver, Jean-Louis Marnat, which he used in his Mini Marcos to achieve eighteenth overall in the Le Mans 24-hour race.

A cold reception

So this was the department Eddie, Gerald and myself were to join. We were no strangers to the Special Tuning lads, having previously worked with many of them. The gaffer, Bill Burrows, was foreman of the Service Repair shop when I was working there during my early years in the factory. What had happened to Competitions could well happen to Special Tuning and the flexible nature of the factory workforce, which was readily transferable from one department to another, had brought about an agreement between the unions and the management. This agreement was to the effect that redundancy would be applied to those employees that had been in the factory the least time, regardless of how long they had been in the department. Little wonder, then, that three long-serving employees were not particularly welcome additions to the staff.

Team Castrol

Despite the closure of Competitions, Brian Culcheth and Johnson Syer, his navigator, were kept on to carry out promotional work for BL International.

There was no budget for a works-prepared car to enter any competitions, but Castrol agreed to pay for entry in a limited number of UK rallies in the name of Team Castro l. The first of these was the 1971 Welsh Rally with an ex-World Cup Triumph. I was not part of the service crew for this event, in which the car suffered from a series of problems culminating in a puncture on a stage, which dropped it to 14th place.

The Morris Marina

The Morris Marina was a much maligned car, but it was not so bad as the press made out. In fact, it seemed to us that the press had not a good word for any of the British Leyland cars, a fact that can't have helped the company when it was struggling in the world markets.

The Marina was designed as a cheap and cheerful fleet car. Before the car was launched Basil was asked by the Marketing Department to prepare one for the RAC Rally, the entry to be made by the BBC Wheelbase programme. The navigator's seat, alongside Brian Culcheth, was occupied - for this event - by Willy Cave, an expert navigator who had enjoyed many success-

Special Tuning service team before the start of the 1971 RAC Rally. L-r: Basil Wales, Michael Steptoe, John West, the author, Bill Burrows and Martin Reade. (Author's collection)

Bill Burrows and Gerald Wiffen servicing in the snow on the RAC Rally. (Author's collection)

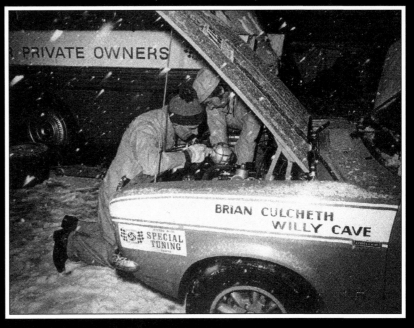

A privately owned car being serviced by
Bill Burrows in wet conditions.
(Courtesy Basil Wales)

es with the BMC team and who was on the production team of the wheelbase programme.

Extensive testing, carried out by Eddie Burnell at the Bagshot testing ground, highlighted the need for a stronger back axle and the shock absorbers were fitted with an extra fluid reservoir, just as the Austin-Healey had been,

so long before. All the body seams were MIG welded. The body was further strengthened by using high density foam to fill the sills, and a substantial sump shield was designed.

The Marina was produced with two different engine size options: 1.3 litre and 1.8 litre. The car used in all the Bagshot testing was the 1.8 option but the 1.3 was used on the rally. This was a deliberate ploy by Basil

and, in consequence Ford, who also used the Bagshot testing facility and had been observing the Marina's progress, did not enter a 1.3 car, which left the Marina with less top class opposition than it would otherwise have had.

This event brought us ex-Competitions men down to earth. Special Tuning was committed to having spares on hand and a crew of mechanics to fit them and carry out any repairs. Fifty per cent of the rally entry was made up of Minis of varying types, from 850s to Cooper Ss. They were scattered throughout the 200 strong entry list with experts having early numbers down to the novices running at the back. It followed, therefore, that a Service Point would have to be manned for an awfully long time.

Our first stop was at 10 o'clock on a freezing cold night. The earlier cars were well prepared but, as we got lower and lower down the list, it became obvious that many of the cars had had scant preparation. The night was spent changing driveshaft couplings, rubber and hydrolastic suspension units, and all the other jobs too awkward for most of the amateur mechanics/ rally crews to tackle for themselves. It was seven the next morning - still dark and cold - before we saw the last of them.

A BL employee, who was enjoying a degree of success in rallying with a 1275cc Mini, had approached Basil with a request for an engine to take part in the event. To Basil's embarrassment, this driver - using a works engine - was beating the works entry Marina on many of the forestry sections. The driver was Russel Brookes, who was to consistently

The Special Tuning 'Wheelbase' 1.3 Marina winning its class on the 1971 RAC Rally.
(Courtesy Basil Wales)

figure in the top ranks of rallying for many years to come. Unfortunately for him, but to Basil's relief, he crashed his Mini, leaving Brian Culcheth to bring the Marina home in eighteenth place, winning his class on its first outing.

Basil's flair for publicity came to the fore when the BBC Wheelbase team was filming. The big British Leyland mobile workshop was in a prominent position in the major service area at Machynleth and Basil had the BBC men do all of their interviewing with the van prominent in the background. Stuart Turner was quick to see the television exposure that BL

was getting and said "If you want to interview me, you'll have to do it inside the hotel!"

Lost tyres

For the 1972 RAC Rally the Marina was fitted with special tyres that could be driven on after a puncture. This was the 'Denovo' tyre, and it required a special rim. The tyre was still in the experimental stage and there were very few made.

I was driving a van with some of the spare Denovo tyres. We got to our appointed place where we were to meet Brian Culcheth for service, early in the morning, after driving and servicing during the night. There were still a couple of hours to wait before Brian would arrive so we settled down to sleep, first booking our servicing space by marking it out with the wheel/tyres. Imagine our horror when we awoke to find these precious experimental tyres gone: somebody had stolen them whilst we were asleep.

Later on in the rally at the top of Sutton Bank in Yorkshire in freezing conditions, the Marina arrived, making painful noises which signalled big end bearing failure. A couple of service crews were on hand for this was a major service area and we soon had the sump shield and sump off. The bearing failure was caused by one of the brass oil orifice inserts in the big end journals coming loose and gouging away the bearing surface. The insert was refitted with a locking compound and a new set of bearings put in. Studded tyres were fitted for the snowy weather conditions and we were able to send Brian on his way again. He had indicated that he would probably need new studded tyres after a couple of stages, which Basil Wales was to carry there for him. To help Basil with the road conditions, the studded tyres were fitted to his Allegro and I was assigned to accompany him to make the change. We had an inauspicious start. Basil had not appreciated the different handling characteristics of the car on studded tyres and we immediately executed a

180 degree spin, finishing alongside the opposite kerb. We proceeded with more caution until descending a winding hill, where the front end of the car slid into the Armco at the side of the road. I was cold and couldn't understand why the heat from the car heater wasn't making any difference. It must have been a nightmare drive for Basil with me wanting the heater full on, added to which, to try to stop my shivering I smoked numerous cigarettes; Basil was a non-smoker. My coldness could be explained by the fact that I had kept on the waterproofs I had been working in and the heat couldn't get through them to warm my body. When we stopped for petrol I took off the waterproofs and felt much better.

It hadn't been possible to clean out the oilways of the Marina engine and, consequently, the debris found its way through and damaged the bearings again. Martin Reade was on hand to fit another new set, but it wasn't long before the problem recurred and Brian was forced to retire.

The Triumph Dolomite Sprint

Triumph had developed a 16 valve engine for the Dolomite. It would later be called the Dolomite Sprint, but, for its first outing, and under considerable secrecy, it was entered in the Scottish Rally of 1972 as an ordinary Dolomite. During the event the car showed signs of a bad rear suspension set-up and we fre-

quently had to change the rear shock absorbers. After a particularly rough stage, Brian Culcheth came in complaining about the rear shockers again and, to our amazement, we found that they had punched up through the top of the turrets built into the body at the sides of the boot space. I had to climb into the boot to weld the turret tops back on. Also in the boot with me was the petrol tank: it was a very shaken Brian Moylan that emerged when the job was completed!

Marina in Finland

In 1974 Martin Reade and I took the Marina across to Finland for the Thousand Lakes Rally. We left from Felixstowe aboard a freighter. There were no other passengers and we messed with the crew, whose meals were of quite good standard. Our interesting voyage took us through the Kiel Canal and along the Baltic Sea to Helsinki.

We made ourselves known to the Leyland Finland people, who were very hospitable. We were booked into a hotel that night and the two Finish mechanics, who were to make up a service crew, showed us the nightlife of Helsinki.

The next day we set out for the rally starting place, 150 kilometres inland at Jyvaskyla. Here, we were to be quartered in student accommodation, with a wafer-thin mattress on a flat wooden bed. It was do-it-yourself accommodation with a laundrette in the basement, but it suited us. The cafeteria served good, cheap meals and had a bar and dance area. Each night a different group would provide music.

When Brian Culcheth and Johnstone Syer arrived, they took up residence in the nearby Ranttisipi Hotel, the rally headquarters, and we occasionally ate there with them.

We had got to Jyvaskyla three days before the rally started and took advantage of this to drive round the route, familiarising ourselves with the location of the service points. As usual on these events, the service point is in a general location. Once there, a good position has to be found, so it was to our advantage that we knew exactly where the best spot would be and headed straight for it before other crews turned up.

The Marina had its greatest successes in Finland on the Thousand Lakes Rally. The main rival was another car with a bad reputation in this country, the Czechoslovakian-built Skoda, but it had proved a formidable rally car, winning its class in many European rallies, driven by the very quick Norwegian, John Haugland. This was Brian Culcheth's main opposition in his class.

The Thousand Lakes Rally is famous for the many blind brows that the cars face on the route. The cars take off on these brows (the Finns call it 'yumping') and come crashing back down some way further on. This is very hard on the suspension and the front of the Marina was soon sagging. There are no front springs on the car, the springing being provided by torsion bars. Every time that we were able to get to the car Brian had us adjusting these bars. There is a limit to the adjustment that can be made before it becomes necessary to take them out and refit them one spline further round; toward the end of the rally we had to do that. Brian would gauge the height of the car by putting his hand between the tyre and the front wing. By letting the car down gently off the jack we were able to give him a great psychological boost when he saw how much we had been able to raise the 'ride-height'. Brian, with his navigator Johnstone Syer, just failed to win his class and was beaten into second place by the Skoda.

The Marina had shown up well on the television coverage and British Leyland Finland were satisfied: they had spent a good deal of money on publicity. One of their Marinas was driven over the stages before the cars were due, throwing out T-shirts printed with the slogan 'The Thousand Lakes with The Morris Marina'. The night the rally ended we had a meal and a few drinks and I retired early to catch up on my lost sleep. I was anxious to start on my way home the next morning, and get there in time to be at my first granddaughter's christening. Martin stayed in the bar chatting and drinking with Johnstone. It was mid-summer and Martin couldn't get over the fact that when he came to bed at one o'clock in the morning it was still daylight. I was sound asleep but was rudely awakened by Martin shaking me "It's one o'clock" he said. I focused my eyes on him, "Have you woken me up in the middle of the night just to tell me the bloody time?" I asked. "Clear off!", or words to that effect!

I travelled home alone, leaving Martin to prepare the rally car for its return journey. When I got back to Helsinki I made enquiries about booking a passage to England, but all I could get was a place on the ferry to Germany, where I would have to book again for the onward passage. I arrived in Germany in the afternoon; a boat would be sailing the next morning. It was fully booked but my name was put at the top of the waiting list. I was determined to get home in time and if I couldn't do it by boat then I was going to drive across Europe. With this in mind, I drove a hundred kilometres on my way, far enough not to waste time but not too far to drive back if a berth was available on the boat. I found an hotel just off the autobahn. The next morning the proprietor 'phoned the docks for me and gave me the news that there was indeed space for my car and me, so the event was rounded off with a pleasant cruise home.

Four wheel drive Mini
The popularity of rallying on television spawned a new event, a

Four wheel drive Mini with modified rear suspension and driveshaft. Note the holes made in the rubber suspension cone to allow softer movement. (Courtesy BMIHT/Rover Group)

cross between racing and auto cross called rallycross. This first appeared on our TV screens in 1969 and BL entered two Minis for their Saloon Car Champions John Handley and John Rhodes. It was a winter event, usually freezing and always wet. The two cars were powered by 1295cc fuel-injected Cooper S engines. The metering unit was mounted just below the dynamo on the weather side of the transverse engine. It had a steel barrel revolving in an aluminium case. The cars were taken to the event - which, to start with, was at the Lydden Hill circuit in Kent - on trailers. On our first visit we tried to start the engine when we got to the circuit and found that the metering unit had seized up. The two dissimilar metals contracted at different degrees in the freezing wind on the trailer. In future, we always took the precaution of pouring hot water over the unit before trying to turn the engine over.

Two private entrants - Brian Chatfield and Jeff Williamson were taken under our wing and the latter soon earned himself the soubriquet 'Jumping Jeff' by the television commentators. These two made up the quartet of what had been the fastest exponents of the sport. But the Championship was won by find of the season, Hugh Wheldon, another Mini driver.

When the Competition Department closed in 1970, Special Tuning was asked by British Leyland International to continue with this high-profile televised event. There were various Mini teams competing successfully and they provided a good shop window for the Special Tuning parts.

The two works drivers were no longer with us, but Hugh Wheldon. David Preece and 'Jumping Jeff' continued to enjoy support from the factory. All this lovely publicity was galling to BL's rivals, who realised that they had somehow to steal the limelight. And they did. Ford's number one works driver appeared on the start line with a Ford Capri equipped with a four wheel drive system that had been developed for the Ferguson PP9, the only four wheel drive car ever to win a Formula 1 race. It was an absolute winner; while the Minis were slithering through the mud, the Capri ploughed on to win by

The Cooper S engine with four Amal carburettors.
(Courtesy BMIHT/Rover Group)

a street. Dutch television also carried the event and Oaf joined in with a four wheel drive car of its own: the event was fast turning into a race between Oaf and Ford.

A four wheel drive car was obviously the thing to have. Triumph had built one which was used to good effect by the Competition Department, But a broken rear suspension arm, in the final rallycross round of 1969, caused the car to crash and be Written -off.

In the MG factory at that time was a prototype military vehicle based on the Moke - the 'Ant' - which was used to tow non-runners off the production lines. Competitions had looked at it but not enough interest was shown by higher management and the project was abandoned before the department was closed.

The call came from Longbridge to build a competitor to the Ford. The 'Ant' was brought into the workshop and the transmission stripped off it. The rear diff unit was mounted on a bracket at the back of a Mini Clubman and driveshafts carried the drive to the wheels mounted on the swinging arms that had been adapted by Bill Burrows and Bill Hine, two ex-Service Department and Special Tuning stalwarts. The gearbox had been modified with an additional pinion wheel driving a propshaft, while the front wheels were driven in the usual way.

One complication was that the additional pinion was in the space normally occupied by a

The steering column gearchange. (Courtesy BMIHT/Rover Group)

Main pic: Testing at the local gravel pit. (Courtesy BMIHT/Rover Group)

John West in tartan tammy with Eddie Burnell and High Wheldon with car 75, and David Preece and Basil Wales at Valkenswaard. (Courtesy Basil Wales)

shaft connected to the gearchange mechanism. This vastly complicated the gearchanging leverage. The first attempt to deal with this entailed attaching a rod to the gearbox which came into the car to a flexible joint carrying the gearlever; unfortunately, the rod worked directly on to the gearbox, missing out the lever that would have been on the missing shaft. The consequence of this was that the gearchange sequence was completely reversed.

The work had been carried out extremely quickly and, subsequently, the car was taken to a big gravel pit where it was put through its paces by Bob Freeborough, who brought along his own Mini as a comparison. Basil Wales, stopwatch in hand, was delighted with the four wheel drive car's improvement over Bob's car, which was modified to works' specification and in which he had had considerable rally success.

The new car was taken to the Lydden Hill race circuit in Kent that weekend, to be driven by Hugh Wheldon on its first public outing. The four wheel drive modification was kept secret until the car was on the start line for the first heat. When the flag dropped, the Mini flew off the line, leaving behind Roger Clark in the Ford, and led for the whole race. In the final heat of the day, the Mini was again in a winning position when the diff pinion sheered.

The next outing was to Cadwell Park, where the rod from the engine to the gearlever broke during practice. The first race was held up while we stripped it out and welded it. When it broke again in the race we put it down to hasty welding, but we had more time before the next heat and took more care; nevertheless it didn't last out the day.

Later investigation showed

Basil Wales, extreme left, with Eddie Burnell and John West at a Valkenswaard rallycross.
(Courtesy Basil Wales)

that the system had serious design faults. I was told to "sort out a gearlever that works". There was a big discussion on how to do this and lots of ideas were chalked out on the workshop floor. The outcome was a steering column control with a lever coming off the gearbox, through a Triumph steering column universal joint, up the column by way of two rose joints to a gearlever standing out from the column just below the steering wheel. This proved ideal and was the final touch needed to make the car a world beater.

The sport was by now international and, once a month, the Mini was taken across to Holland where the principal venue was a race circuit in Valkenswaard. Very soon the continental car producers latched on to the publicity potential and four wheel drive cars were produced with engines twice the size of that used in the Mini. Eventually the sport organisers decided that the big money being spent by the factories was making the sport non-viable for the club driver and four wheel drive was banned. But the little Mini has again made its mark in international motorsport.

Basil Wales moves on

Two major management changes took place in 1974. Bill Price, who had left the factory in 1970 with the closure of the Competitions Department, was brought back in to take up the new position of Workshop Supervisor. To us ex-Comps men this was a welcome addition to the staff, because, all through the Competitions days, although Bill was assistant to various managers, he had always had a good rapport with the shop floor and was treated as one of the lads.

The other change came about when Basil was offered an important job with Unipart at Cowley. His place was filled by Richard Seth-Smith, a Leyland PR man.

5 Richard Seth-Smith

1974 1000 Lakes

Following the 1973 class result of second place, BL and Brian Culcheth were anxious to have another go at the Skoda and, again, I went as part of the service team, this time accompanied by John West.

After the first stage the Marina broke the sump shield mountings. Luckily, this happened within easy reach of a BL workshop, which was on a road section of the rally route. Time taken to effect the repair would have to be made up before the next control ten miles away. The mountings were completely broken up and, with time flying by, Brian said "Weld the bloody thing to the chassis frame". There was no quick alternative, so that's what we did and, with eight minutes to go, Brian hurtled out of the workshop, breaking every speed limit on the way and made the control with no time to spare. Just beyond the control there was a speed trap, had it been just before the control the Marina

Richard Seth-Smith

would undoubtedly have been disqualified.

John and I made for our next rendezvous, accompanied by two Finnish mechanics. On the way we stopped at a filling station to take on fuel for ourselves and the rally car. While I was at the counter paying, two men came to stand one either side of me "You buy some of our alcohol"

one of them said. I declined this offer and, having paid for the petrol, went into the toilet, where they followed me. One of them produced a knife and stood behind me. Just then John came in. "I've got trouble here" I said "go and get the rest of the lads". At this, the pair made off, much to my relief

The Marina went on to win its class, which was hailed as a great achievement for a non-Scandinavian driver to beat the locals on their own ground.

When we got the car back to England we found that, in welding the sump shield to the frame without the benefit of a rubber mounting, the pounding had all been taken by the chassis frame and had effectively written the car off, which didn't please our foreman, Bill Burrows, one little bit. "That was a bloody daft thing to do" he said, and we could not but agree with him.

There was, however, a fortuitous sequel to this, as far as I was concerned. The frame was

Left: The author in the ex-works Marina, getting advice from Martin Reade at an autocross. (Author photo)

patched up and the car used for experimental purposes. In due time it was pensioned off. I was, at that time, secretary of the factory motor club and was able to buy the Marina complete with an 1800cc engine and Triumph Dolomite overdrive gearbox for use by club members for the nominal sum of £1.00. We had some great times autocrossing and rallying with it, until two members using it on a national rally put it into a ditch, causing too much damage for us to want to repair it. It was sold with a healthy profit to the club's finances.

The author commentating on an MG Works Auto Club Autotest. (Author photo)

Gerald Wiffen, on ladder, and the author with the Sherpa service vans. (Author photo)

Leyland ST
Richard Seth-Smith brought his public relations skills with him to Abingdon and the new image was epitomised by the department's new title of 'Leyland ST'.

However, the new image didn't bring about much of a revival of the department's fortunes: although some class wins were achieved with Marinas, Allegros and Dolomite Sprints, these were hardly the cars in which outright wins were achieved.

The Triumph TR7
1974 brought the car it was hoped would be a worthwhile competitor in international rally-

Left: The author autotesting with an ex-works 'Barge'. (Author photo)

143

ing, the Triumph TR7. But it was plain from the outset that a lot of development was required and it was not until the Welsh Rally of 1976 that the car made its first appearance.

Rally *aficionados* still remembered, with affection, the Austin-Healeys that stirred the blood so many years before, and the belief that had been fostered by BL that the TR7 was its descendant, so its performance was anticipated with great optimism.

I was in one of the Sherpa vans that we were now using as service vehicles and which were decked out in similar livery to the rally cars. The vans had been extensively modified to our own specification. The front drum brakes were removed and discs were fitted, the mounting brackets for the callipers were handmade in the department. Stronger springs and shocker absorbers were fitted.

We used many compressed air tools in the workshop - ratchets, chisels, nibblers, drills and wheelnut guns - which greatly speeded up the work. These were sorely missed when working out in the 'field' and Bill Burrows came up with the idea that we should have an air tank filled by a small MGB air-conditioning compressor. He gave me the job of putting it all together. The compressor was mounted at the front of the engine and driven by the fan belt. A pipe carried the air to the tank via a blow-off valve set at the same pressure as that used in the factory. The tank was

slung under the body quite well up between the chassis side members. From the tank, a pipe led to twin bayonet fitting outlets at the back of the van. This seemed to work quite well but after the first run we found that, on both the Sherpas so equipped, the big end bearing on the compressor units had disintegrated. In discussion with the design office, we came to the conclusion that the pumps were having to operate at full pressure all the time that we were travelling and were not up to it. The remedy was to fit an electric clutch to the driving pulley which we switched on as we were nearing our service point to ensure that pressure in the tank was immediately available, and switching off to allow the pulley to freewheel while we were travelling any appreciable distance.

Our two Sherpas were parked about a mile from the start of the rally in Cardiff and the streets were thronged with rally enthusiasts. Their cheers grew loud as the TR7s came in sight and many of them came to us and said how great it was to have a truly British manufacturer back in international rallying. Alas, the euphoria was shortlived, for after the first stage Brian Culcheth came in with his car blowing out clouds of steam: its cylinder head gasket had blown. The engine was the same 'slant-four' used in the Dolomite Sprint and this problem had been experienced by a team that raced a Sprint. Stronger head studs had been made so

that they could be tightened down to a higher torque setting. Tony Pond's car also expired, with its cylinder block cracked at the centre main crankshaft bearing. This had also been encountered on the race cars but the information had failed to reach Leyland ST.

In addition to engine problems, other faults started coming to light. The rear axle was located by parallel lower links and diagonal upper links. These upper links ran from a mounting in front of the axle at the outer chassis members, across to mountings on the axle tubes, where they were fixed to the differential housing. The arc described by these links as the axle bounced up and down took the axle end of the links very slightly apart from each other, which was taken care of by using big soft rubber bushes. The axle being located by these soft bushes was absolutely no good for competition work and harder ones were made, but these didn't stop the link ends from moving away from each other and they literally pulled the axle apart.

The diagonal top links were discarded in favour of parallel links and, to compensate for the loss of lateral location, a panhard rod was incorporated. This is a rod mounted from the chassis, behind and in line with the centre of the axle to a fixing point on the axle on the other side of the car; not the ideal solution but the best compromise possible.

The front suspension also came in for some criticism for a phenomenon known as 'bump

RAC Rally, 1979, Simo Lampinen's TR7 getting the treatment at a Leyland ST Service Point. (Courtesy Den Green)

steer'. This is caused by incorrect steering geometry which means the car tends to steer itself when leaning heavily into a corner. Because of the design of the suspension, this fault proved incurable, although extensive modification brought about some improvement.

Bill Price

In 1964 I turned down the opportunity to take over as manager of the BMC Special Tuning Department as I was keen to stay in the Competitions Department at the 'sharp end'. I left the MG Car Company in 1970 after the closure of the BL Competitions Department but kept in touch, through my friend, Brian Culcheth, with events as they unfolded within the BL empire. After an interview with Keith Hopkins and Simon Pearson, in 1974 I was offered the post of Workshop Supervisor in the Special Tuning Department. Ever since leaving in 1970 I had dreamed of the possibility of getting back to the job I loved. The long-term plan was to launch a rally programme, probably in 1976.

I realised at the time that my appointment would probably not be welcomed by the staff at Abingdon, mainly because there were almost certainly already employees with similar qualifications. In the early 1970s, thousands of people were unemployed and the

Bill Price.

unions in BL were pressing for job vacancies to be advertised internally and filled where possible from within. I am still not sure how I managed to get back!

Later in 1974, Richard Seth-Smith was appointed manager of the relaunched Leyland ST, taking over from Basil Wales who moved to Cowley. It was a very difficult time, with financial and political problems taking up too many working hours. Early in 1976 Richard moved to Redditch as Product Affairs Manager, retaining control of Leyland ST, and I was very surprised to be offered his job as manager of Leyland ST. I was confident that with the commercial side in the capable hands of John Kerswill, I would be able to concentrate on the build-up and co-ordination of the planned rally programme.

As history records, Leyland Cars launched a new rally programme in 1976, 'spearheaded' by the Triumph TR7. Unfortunately, the press rather exaggerated the potential of the new car and the first few events in 1976 ended in disaster, but that is another story. Nevertheless, I never regretted going back.

Bill Price

TR7 gearbox modifications

Richard Seth-Smith moved back to Leyland PR work and his place as Leyland ST Manager was

Bill Price with the team of Special Tuning mechanics. (Author photo)

Eddie Burnell (left) and Gerald Wiffen
with a TR7 engine on the test bed.
(Courtesy Oxford Central Library)

taken by Bill Price. To replace Bill as Workshop Supervisor, Den Green, the ex-Competitions department foreman, was brought back in from the production line where he had been languishing since Competitions closed.

Development work continued on the TR7 and it gradually evolved into a competitive rally car. A great step forward would be made when the V8 engine was incorporated.

The department had now acquired a rolling road and engine test facility. Eddie Burnell and Gerald Wiffen were enlisted as engine builders and performance improved. The weak link now was the gearbox.

The Mille Pistes

The *Mille Pistes* was a single venue event. The roads across a French military training ground in Provence were circuited four times. It was an extremely rough route and was destined to take its toll on the majority of the entry.

Practice was not allowed but the contestants were led round the course by the Course Car, which, on this occasion, was our Range-Rover that the rally authorities borrowed for the job, complete with driver Bill Price.

After the familiarisation circuit, Tony Pond was concerned that his gearbox was showing early signs of trouble. The gearboxes having been a source of trouble on previous events made the drivers very sensitive to any sign of malfunction. With the car jacked up on axle stands as high as possible, Martin Reade and I, working from under the car in the dust and rubble, carried out the awkward job of changing the box. The large bell housing made withdrawal of the box awkward but the real difficulty was experienced when we tried to refit the new one.

The end of the first motion shaft of the gearbox has to locate into the centre of the flywheel and the locating recess has a bush or bearing for the shaft to fit into. On this car the bearing was not a whole unit but a ring of needle rollers each 15mm long and little more than a millimetre in diameter. These needle rollers had to be around the inside of the locating recess, held in place by grease, a simple enough operation with the engine and gearbox on a bench in a workshop, but trying to locate the end of the shaft into the ring of rollers without dislodging them while manipulating the heavy gearbox, lying on our backs, proved a long and frustrating exercise.

Had this occurred in the middle of a rally it would have undoubtedly brought about the car's retirement. Work was put in hand when we returned to Abingdon to ease the problem of gearbox changing. A caged roller assembly was fitted into the flywheel instead of the loose needles. Martin Reade was given the task of easing the operation, by making the gearbox detachable from the unwieldy bell housing. With this done the box was secured to the housing by a ring of Allen screws which we could

Brian Culcheth making a retirement presentation to Bill Burrows and his wife. (Author photo)

rapidly remove replace; with the aid of our air-operated tools, it became feasible to change a box during an event.

We were able to take advantage of the new gearbox arrangement on a later RAC Rally. Our Sherpa vans were now equipped with VHF radios. I was teamed up with Les Lloyd, a new recruit to the staff who was a successful motorcycle trials rider, Les spent his lunch hours in the factory making a Unicycle, which,

when completed, he would ride round the factory to everybody's amusement.

We were travelling between service points when over the radio came a call from Pat Ryan, one of the TR7 drivers, reporting that his gearbox had failed on the M6 motorway. We listened to the conversation he had with Bill Price and it became obvious that we were the nearest service vehicle. Les consulted the map. From Pat's description they were about a mile from a Motorway Service Area, where it would be possible to carry out the gearbox change. He gave Pat this information and told him to try to limp as near to the Service Area as he could. We caught up to the TR7 about half a mile from its target where it was stopped, unable to continue any further. Contrary to rally regulations, we quickly hitched a tow rope on and towed the car the half mile to where we could work on the car. The job was accomplished in twenty minutes, including removal of the full-length undershield. This was a great improvement on other attempts, thanks to Martin's modification.

7 *John Davenport*

In 1977 John Davenport was appointed Grand Supremo of all motorsport activities carried out by British Leyland, which, besides the rally programme, were racing competitions and championships. Bill Price remained as Leyland ST Manager.

With Davenport came Dave Wood with the title of Engineering Liaison Manager.

Catering

Dave's wife also played a part. She took over responsibility for catering in the caravan acquired for the job. No longer were we mechanics required to have a pot of tea or coffee ready for when the crews arrived for service, or open a tin of chicken supreme. Our homely catering efforts had been a source of fun in the earlier lighthearted days, and the crews would award us stars for the level of catering provided. My highest award came when I was servicing near a take-away stall and provided a supper of chicken and chips for all of the crews.

The caravan was first used in the 1987 RAC Rally with the tow car being driven by the workshop "Gopher" Andy Mansell. The caravan was also regarded as a convenient method of transporting heavy spares to the main service areas, where the crews would theoretically have time to avail themselves of the meals. Andy took a short cut over a yellow road in Wales,

John Davenport

which wasn't a good idea as the road wound its way over a mountain and part-way up the car refused to pull the heavily-laden caravan any further. They had reached a conveniently wide stretch of land and Andy tried to turn the caravan round, getting stuck halfway with the back end of the caravan over the edge of the road. "What are you going to do now Andy?" asked Dave's wife, "Panic!" he replied. But, in fact, he didn't. He managed to complete the turn and descend to a less steep part of the road where they took off half of the load and left it by the roadside, drove to the top of the hill and unloaded the other half and then came back down for the remainder.

All of this took time and they were late arriving at the control where the ravenous crews were expecting a hot meal.

Tour de Corse recce

In October of 1978, I was dispatched to Corsica with a TR7

in the back of a Leyland Terrier truck. The TR7 was to be used by Tony Pond and Fred Gallagher to recce the stages on the Tour of Corsica Rally.

I had a short run in the Terrier to familiarise myself with it during the week before I set off, but it had been some time since I had driven anything bigger than a Sherpa van. Taking off to drive through France with a truck laden with the car, a spare axle and gearbox, jack, tools and all the rest of the equipment needed to cover as many of the eventualities as possible, was a different matter.

Once out on the road I soon became acclimatised and fancied myself a genuine trucker. Arriving at Portsmouth, where I was due to embark for the hop across the Channel, I joined the queue of trucks for the booking-in office, where I was told that my ticket was for a private car and I was in the wrong queue. Feeling very deflated I took my place among the other lesser beings.

Once in France I quickly adjusted to driving on the right and the 60mph speed limit. The route that I had opted for was to take me to Paris on the A 13, then round the *Peripherique* to take off down the A6 and A 7 to Marseille. Nearing Paris I spotted a sign with a truck on it and an arrow which I assumed instructed me to leave the *autoroute* and rejoin it further on. I failed to find another signpost to take me back on route and drove to Paris on the N road. At this stage I wasn't concerned;

after all, the *Peripherique* circled Paris and I was bound to hit it sooner or later. But I didn't and drew even nearer to the city where I knew trucks were banned from the centre. I could see the Eiffel Tower and took the next likely looking road away from it, but the Tower loomed larger no matter how I tried to avoid it. Eventually, and with great relief, I hit the inner ring road and picked up my bearings again.

The next morning I was in Marseille and went to the shipping office to confirm my booking on the ferry for Corsica, only to be told that they could find no evidence of a booking having been made. However, there was room for me and the Terrier on a Ferry devoted entirely to trucks.

When I got on board I found my cabin, which I was sharing with another trucker, who I never saw. A very palatable evening meal of meat, potato and beans, followed by cheese, was served in the mess room. We were six to a table and the attendant put two bottles of wine at each table; every time a bottle was emptied he cheerfully replaced it. The cost was included in the ticket.

I duly did my share of emptying the bottles and found my way to bed. When I woke the next morning my cabin mate was up and gone for his breakfast, which was bread and jam washed down with strong black coffee.

It was a short drive from the docks to the hotel where I was to join up with Tony Pond and his navigator, Fred Gallagher, who

had driven down in another TR7. The hotel had a large garden with a concrete area under some trees where I could comfortably work during the hot days. I was thankful for this shade when, on the second day of their recceing, Tony and Fred came back with a broken axle for me to replace. I had the help of the enthusiastic hotel waiter to lift the heavy axle over the awkward bits.

After a couple of days working on the northern end of the island we moved South to Ajaccio. Tony told me that he would want me to meet them the next day in the mountains some way across to Porto Vecchio. The map showed it to be a very narrow and twisty road, and the reality was no better. Driving a right-hand drive truck on the right side of the road proved to have its advantages, when, with the drop on my side, I met buses and lorries coming the other way. I was able to look out of my window and see how close to the verge I was while we edged past each other.

Not so lucky was the driver of a road roller which had apparently broken down and was being towed up the hill. The tow lorry was going too fast for him and he was having extreme difficulty keeping to the winding road; one moment heading for the drop and furiously winding the steering wheel, the next heading for the mountain side, each time leaving go of the steering to gesticulate and swear at the towing driver. This was so funny that the truck driver couldn't continue but had to get out of his cab and lean against his front wing, weak with laughter.

I spent two weeks driving round the island, some days staying at the hotel in Ajaccio to repair the second recce car, but whenever I travelled it was with the spare car and all the tools. The car was loaded into the Terrier via two ramps and pulled by an electric winch. After the car and the ramps were loaded I would lift the jack in. At the end of one trip I opened the back of the truck to find the TR7 almost touching the roof: I had put the jack under the car with the handle still in place and the undulating roads had caused the jack handle to move up and down, lifting the car as I drove along.

I flew back to England at the end of the fortnight and, for the next fortnight, my place was taken by Les Lloyd. But all this recceing and testing came to naught, for soon after the start of the rally both cars lost the oil from the gearboxes due to loose drain plugs. This mysterious happening revived memories of the 1967 event, when all three Minis suffered from overheating due to loose fan-belts, raising suspicions of sabotage.

My reluctant departure

Early in 1979 I received an offer from the Service Manager of Morris Garages in Oxford to manage a small satellite garage at the southern part of the city. The garage had a small workshop with three mechanics, plus a showroom and two salesmen. It was situated adjacent to a Park and Ride area. I was now in my forty-ninth year, twenty of which had been spent in long hours of competition car preparation in the workshop, followed by repairing rally cars in all conditions at all hours of the day and night. The novelty was wearing off: the latest management team didn't have the same rapport that we had enjoyed in the past; the car wasn't much of a joy to work on; I missed the boat for promotion earlier on and now the younger, qualified ex-apprentices had gained considerable experience, so I couldn't see any way forward for me. The biggest wrench was going to be saying goodbye to Bill Price, Tommy Wellman, Den Green and all the other lads with whom I had shared so many adventures. I visited the garage a couple of times and, after a lot of heart searching, decided to take the job offered by Morris Garages.

The Morris Garages Service Manager, Alan Best, agreed that I should start my new job after I had been on one more rally - the Circuit of Ireland.

The TR7 was now equipped with the V8 engine and was called the TR8. The latest modification was dry sump lubrication. A belt-driven pump was mounted on the front of the engine to pump oil from a tank in the boot into the engine oilways. The scavenge side of the pump took the oil back to the tank. This was far from 100 per cent effective; sometimes the engine filled with oil when the scavenge side failed. Other times the engine was starved of oil when the pump side failed. This was what put out our last hope on the Circuit of Ireland: two of our three car entry suffered accidents, causing retirement, and the engine on the third one fell victim to the failed dry sump lubrication system.

My last event, therefore, was not the cause of any great celebration. Back in the factory my remaining days were the most miserable that I have ever spent, and many times I was on the brink of calling the whole thing off. The day before I took my leave, Bill called me into his office. I half hoped that he was going to talk me out of going, but it was to talk about the times that we had had and to wish me luck in my new job.

And so I had serviced my last rally - well, almost. Later that year I went to watch the RAC Rally and, on one occasion, was at the control at the end of a stage when a TR8 with a puncture on the nearside rear wheel arrived in a hail of shingle and rubber smoke. Fred Gallagher leapt out to change the wheel and was amazed when a spectator got the spare out while he was getting the old wheel off, put the punctured tyre away as he tightened the new one, then neatly stowed the jack and wheel brace. It was the middle of the night and he was in too much of a hurry to recognise me, but when I saw him later and told him about it, he said he wondered how a stranger knew where everything went.

I had one more close encounter when I volunteered to marshal on the Blenheim Palace stage of an RAC rally. My job was on the gate where the competitors went on to the estate. Only competitors were allowed through but Bill Price arrived in a sweat; Tony Pond had smashed his windscreen and, as a temporary measure, Bill had to get some motorcycle visors for Tony and Fred to wear. I was dealing with another car when he went through.

The MG factory was closed in 1981. Leyland ST continued for a short time after, in the Morris factory in Oxford. But many of members of the staff, including Bill Price, were made redundant.

But what memories!

Appendix
Service Crew Instructions

This Appendix illustrates the complex timetable that the competition managers - with the aid of the navigators who had reconnoitred the event routes - had to compile, not only so that the service crews knew where they were expected to be and how long they had to get there, but also to ensure that the correct fuel and tyres were on hand when required.

The organisation necessary often turned into a rally within a rally, with us mechanics as competitors, leapfrogging each other to form a network of support over whole countries, continents and, in some cases, halfway around the world.

The dates of travel out and return on this schedule give an indication of the length of this event, which started on the 18th of September and finished on the 26th. It was the longest event in the European calendar.

BRITISH LEYLAND MOTOR CORPORATION

Competitions Department

ABINGDON-ON-THAMES, BERKSHIRE

TELEPHONE: ABINGDON 251   TELEGRAMS: EMGEE, ABINGDON   TELEX: 83128

TOUR DE FRANCE 1969

Revised Travel Schedule

Saturday 13th September

    Barges, JBL 496D, JBL 492D and NBL 128E with Triumph UJB 643G and Rally Minis URX 550G, URX 560G and OBL 45F with D. Watts, J. Evans, T. Wellman, D. Plummer, B. Meylan, R. Vokins, R. Brown, D. Argyle, W. Price, D. Pike and J. Syer drive to Southampton, report 23.30 hrs., depart 23.59 hrs. for Le Havre., arrive 07.00 hrs. on Sunday 14th September.

Sunday 14th September

    Above party drive to Paris (Minis on tow bars in City traffic) to report at Charolais Goods Depot between 15.00 and 18.00 hrs. Car Sleeper departs Paris at 19.15 hrs. arrive Nice at 06.45 hrs. on Monday 15th September.

Monday 15th September

    P. Browning, P. Hopkirk, A. Nash, B. Culcheth, J. Handley and P. Easter fly London to Nice, Flight AF 950, depart 10.20, arrives 12.20. Taxis to Hotel.

TRAVEL HOME

Sunday 28th September

    Party drive to Le Havre to cross Le Havre - Southampton on Monday 29th September report 22.00 hrs., depart 23.00 hrs. arrives Southampton 07.00 hrs. on Tuesday 30th September.

The instructions to mechanics showed meticulous attention to detail, and covered every possible contingency.

R.A.C. RALLY 1965
Mechanics' Instructions

1. Read the general instructions carefully, many of the points concern you.

2. It is essential that at least <u>one</u> member of each service crew is at the meeting at 9 a.m. on Sunday, together with all maps. Diana will be at this meeting and will take note of the map references to telephone Geoff Halliwell of Tillotson's on Monday. She will also take note of instructions re Porlock tyres for H. Liddon's friends. D. Watts to be at the meeting to determine final service points between Hereford and the finish, for him and C. Humphries. If things are going well they should plan to wash cars at this point.

3. You have 1" maps for tricky areas, Esso maps to cover the whole route and an RAC route between major places nearest to your service points. The RAC routes are for very general guidance only. Study the 1" maps before Sunday (care - many of them have old markings on them!)

4. There is a mistake on the general instructions for the Perth hotel: for Green (single) and Whittington/Vokins (double) <u>read</u> Whittington (single) and Evans/Brown (double). No rooms have been booked in Perth for Crew C (Bradford/Poole) because if all is well with our cars they can head south before sleeping. We have not been able to find rooms for Crew I in Perth but there will inevitably be cancellations among the rally.

5. <u>All</u> crews who are not at Perth must telephone the Salutation (tel. No. 22166) on Tuesday evening. No rooms have been booked for crews not at Perth because they can decide where to stay themselves on their run to their first service point of the second half of the rally.

6. Study the modified Road Book before Sunday in case you have any particular queries in connection with your area.

7. Unless there is bad fog cars should have plenty of time at all controls so service them before they clock in.

8. Leave <u>immediately</u> after Car 44 (except Crew I).

9. There is no order of priority among our runners although one may develop as the rally goes on. Treat Jones and Fall as works cars as they are in our teams.

10. Article 51 reads "No work of replenishment of any kind may take place on a car during the period that it is within a Control area, other than replacement of a wheel with a deflated tyre and/or the cleaning of lamp glasses, windscreen and windows. A "Parc Ferme" shall, for the purposes of this regulation be deemed a Control. The Breaking of any seal specified in Article 18, other than by an Official appointed to do this duty, will be deemed a breach of this Regulation." (At many controls of course there will be official service areas.) The seals referred to cover "sump, gearbox differential, differential cover". The drivers and/or co-drivers should be aware of the location of these seals.

11. The RAC route near Wells appears to go a long way round for some crews (e.g. to Axbridge). Check with W.R. Price before the weekend as the road works causing this diversion may be cleared before the rally.

12. The service sheet for the person listed first against each car has inked against each point the distance and time to get to their next point.

RALLY H.Q. SKYPORT 6611.— 9522.—1~2~75 11
SUTTON. COURTENY. 295 EXCEPT 8-835 E
455 · 530 AM
B. MOYLAN.

055668 F.N. MECH.

BMC Service - R.A.C. Rally 1965

1. Service crews will be in position by times shown.

2. Except for crew "I" all service cars should leave immediately No. 44 has gone through. Crew "I" should stay in place for the whole rally.

3. The exact location of many service points (those marked *) will not be determined until the exact road section is known; it may then be possible to cover cars twice from one point. Service crews to meet at Abingdon at 09.00 on Sunday to finalise positions.

Time	Place	Crew	Comments
Sun.21st	LONDON	G (E)	May NOT be necessary for you to be at meeting at Abingdon. Can decide in London.
14.15	169/791580 (end of S1)	H	Also near to start of S1 · also 167/190445
15.15	167/188447 (end of S2)	A	Also near to start of S2
15.30	167/092½461½ (end of S3)	C	
16.20	166/755371 (end of S4)	D	
16.30	M2. Camel Hill Cafe.	B, F, I	Large area for service at cafe. F leaves immediately after 44.
18.45	* S5 (map 164)	G	161 miles / 5¾ hrs. 976423. Racing tyres there by courtesy of
19.15	* S6 (Porlock - 164)	(E)	Friends of H. Liddon, Esq. 855863. Twice!! Nº24 B. WILLIAMS LEND TYRES IF AMIABLE.
20.15	* S7 (map 164)	F	
21.00	M3. Bristol Airport	A, C	Service area will be controlled by officials. Body repair equipment & experts supplied by G. Mabbs, Esq.
Mon.22nd			
00.15	* S8 (maps 155/143) 639092.	G	
00.45	155/554986 (end of S9)	D	
01.30	M4. Abergavenny ROSLAN→A40	B, H, I	Service space may be limited here.
03.30	141/886413 (end S10)	(E)	95 miles / 3¼ hrs.
04.30	* S11 (map 127) 693696	G	
04.30	M5/6. Devils Bridge	A	Reasonable service space, controlled by officials.
06.45	* S12 (map 127) 755793.	G	
07.00	* S13 (map 127) 688956	F	
07.30	* S14 (maps 127/116) 867101 ABERANGEL	D	
08.30	* S15 (map 117) 014175	(E)	272 miles / 20 hrs!
10.30	M7/8 Oulton Park	B, H, I	Ample space. Line Transporter & fork lift up with any TV cameras.
-	Between M8 & M9. 95/826487	-	Mr. Holden of Gisburn Garage (on left entering town) will assist if necessary and has numbers of works cars. No petrol on credit!
16.00	M9. THIRSK GOLDEN FLEECE SKIPTON →461	C	+ J. Organ dropped off (& collected) by A. There is NO room for service at Golden Fleece Hotel.
17.30	* S17/S18 (maps 91 & 92) 5.9842	A	
18.30	* S19 - S23 94/749911) TWICE 93/894976	G	+ G. Halliwell Esq. & mech. from Tillotsons in yellow van (with welding)
20.45	M10 CHARLTONS 86/640154	C	BIRKBROW SERVICE STN A171 NEWGATN →

The six-figure reference on a one inch ordinance survey map pin-pointed the exact location of a service point.

INDIVIDUAL MECHANICS INSTRUCTIONS
B. MOYLAN

With Rolls (JBL 492D) and Wiffen and Bradford cross Southampton -
Le Havre on Tuesday, 14th April, depart 22.30 hours, arrive 07.00 hours,
on Wednesday, 15th April with the following:- Rolls NBL 128E (Pike/
Whittington/West), Castrol 1800 RME 522F (Stacey/Plummer/Bisp), Land
Rover MXC 337H (Hall/Hogan) and Triumph VBL 197H (Wood/Reade).   Drive to
Titograd (1500 kms).

Stay night Monday, 20th April at Hotel Crna Goro, Titograd.

First car due 16.48 hours on Tuesday, 21st April.   Control closes
23.55 hours.

Service after control at bridge before Titograd main road.

Watt,s Green and Vokins, Wood and Reade will arrive by separate schedule.

Stay night Tuesday, 21st April at Hotel Crna Goro, Titograd.

On Wednesday, 22nd April return barge to Titograd airport for collection
by Plummer and Stacey from Pec and fly Titograd to Nice with Watts, Wiffen,
Bradford, Bisp as follows:-

Titograd - Belgrade, depart 07.30 hours, arrive 08.10 hours, Flight JU 749
Belgrade - Rome, depart 14.20 hours, arrive 15.50 hours, Flight AZ 511
Rome - Nice, depart 16.45 hours, arrove 17.50 hours, Flight AF 623.

At Nice collect hire car (booked in you name) and drive to Camporosso
(80 kms) with Bisp and Bradford where you will meet Wellman and Eales in
3 litre (TMO 938G).   Vokins will also arrive by separate schedule.

First car due Camporosso 15.00 hours on Thursday, 23rd April.   Control
closes 23.05 hours.

Service after the control at the end of the prime.

After the control closes, leave Bisp and Bradford to return home with
3 litre and taking Wellman and Eales with you, return to Nice in hire car and
fly Nice to Lisbon on Friday, 24th April with Watts, Wiffen, Wellman, Elkins,
Browning and Zafer as follows:-

Nice to Lisbon, depart 10.35 hours, arrive 13.55 hours, Flight PA 155.

Collect service car from Goncalves and emergency kit from transporter
and drive with Wellman to Pampilhosa (200 kms) to service at the end of Prime
5.

First car due 03.45 hours on Saturday, 25th April.   Control closes
14.10 hours.

Service after the control at the end of the prime.

Return to Lisbon after the control closes.

Stay at Hotel Nau, Cascais, Lisbon.

The European leg of the London to Mexico World Cup Rally of 1970: a preliminary to the main event in South America!

INDIVIDUAL MECHANICS INSTRUCTIONS
B. MOYLAN/J. EVANS

On Monday, 27th April fly Lisbon to Rio on charter Britannia departing 09.00 hours, arriving 23.15 hours.

Disembark at Rio with Watts, Vokins, Wiffen, Reade, West and Adams (Dunlop). The spares and tyres for all the Brazilian service points will also be unloaded at Rio.

Stay at Hotel Gloria, Rio until Monday, 4th May. (See D. Watts instructions concerning the handling of spares, allocation of tyres and service vehicles by Rio team).

On Monday, 4th May go with local driver in truck with spares and tyres to Canella (end of Prime 8) approx 700 miles.

First car due 17.00 hours on Saturday, 9th May. Control closes 03.00 hours.

Service after the control at the end of the prime.

After the control closes drive to Porto Allegre (50 miles) and fly Porto Allegre – Buenos Aires – Santiago on Sunday 10th May as follows:-

Porto Allegre – Buenos Aires, depart 18.30 hrs, arrive 20.05 hrs, flight SC 109
Buenos Aires to Santiago, depart 05.00 hours, arrive 08.55 hours, flight AR 526
Stay overnight in Buenos Aires (City Hotel).

At Santiago you will be met by representative of Leyland Chile who will provide truck and local driver with prime kit.   Drive to Puyehue (600 miles) and service as close to the Chilean/Argentine border as possible.   Prime 11 finishes about 30 miles on the Argentine side of the border.

First car due 15.00 hours on Tuesday, 12th May.   Control closes 22.30 hours.

After the control closes return to Santiago.   Stay night 13th in Santiago.

Fly with Green and Browning in Cessna via Mendoza for customs clearance and onto San Juan departing 07.00 hours, arriving 10.00 hours on Thursday, 14th May.

Take a taxi/hire car to Rodeo (150 miles) and service before the start of the prime.

First car due 20.00 hours on Thursday, 14th May. Control closes 07.00 hours.

After the passage of the works cars return to San Juan and fly in Cessna to Tinogasta, departing 00.30 hours on Friday, 15th May, arriving 02.30 hours.   This is in the middle of prime 14 and will be an emergency stop only.

After the passage of the works cars fly to Salta in the Cessna arriving 09.00 hours on Friday, 15th May.

Here you will meet up with Wiffen and truck with control kit which has come from Saladillo.

Moylan/Evans (1)

The London to Mexico World Cup Rally, 1970. These instructions gave no hint of the adventures we would experience before reaching the end of the page!

Index

Works Rally Mechanic